P9-DMV-692

100 FLOWERS

And How They Got Their Names

100 FLOWERS

And How They Got Their Names

DIANA WELLS

Illustrated by

Ippy Patterson

Special Edition for PAST TIMES®, Oxford, England

Published by

ALGONQUIN BOOKS OF CHAPEL HILL

Post Office Box 2225

Chapel Hill, North Carolina 27515-2225

a division of

WORKMAN PUBLISHING

708 Broadway

New York, New York 10003

Printed in the United States of America.

Published simultaneously in Canada by
Thomas Allen & Son Limited.

Library of Congress Cataloging-in-Publication Data

Wells, Diana, 1940–

100 flowers and how they got their names / Diana Wells; illustrated by Ippy Patterson.

p. cm.

Includes index.

ISBN 1-56512-138-4

1. Flowers — Nomenclature (Popular) 2. Plant names, Popular. 3. Flowers — Folklore. I. Title.

QK13.W46 1997

582.13'014 — dc20 96–22296

CIP

PAST TIMES®

For my sister Sheila (1936–1995)

and her nephew, my darling son, Quin (1971–1995)

CONTENTS

ACKNOWLEDGMENTS

Thanking family and friends would be as superfluous as thanking peristalsis, the essentiality of which I take for granted, but there are some whom I would particularly like to thank. Frances Greene, Janet Evans, Ellen Fallon, and all the other librarians whom I pestered mercilessly for information and seemingly unobtainable books, which they obtained and I grumpily returned, weeks after they were due. Betsy Amster and Angela Miller, my agents. Elisabeth Scharlatt, Robert Rubin, Amy Ryan, and Tammi Brooks for their skill and encouragement. Pat Stone and the readers of *Greenprints* for their heart-warming enthusiasm. Dr. Candido Rodriguez Alfageme and Dr. Erik A. Mennega for invaluable assistance. Dr. Peg Stevens for her gentle and unfailing help and kindness. Gratitude also to my word processor, so hated at first but finally respected if not loved, even though it never did give me back those pages that disappeared.

INTRODUCTION

We do not read of flowers in the Garden of Eden, but of trees—trees that (except for one) were given to us as food. Nevertheless, those of us who plant flowers have, perhaps, a sneaky longing for Eden, made for our delight, a garden in which Adam was allowed to give names to everything. To name is to possess, as conquerors know. Or so we might wish.

As to when we first became aware of plants not essential for food, the Old Testament doesn't help much, but it must have been early on. Some plant names go back to before we have records, when flowers were used for charms and protection; their names are the stuff of myths, answering our deepest fears and longings, our earliest whimpers in the dark for comfort. The Greek gods, we are told, usually to preserve love (love being what we most crave), had the power to turn humans into plants so they would not die. So it is that Daphne and Hyacinth and Narcissus, and all the poignancy of their loves, are still with us in our gardens.

Other flower names go back to the fear of illness and the mystery of healing, even if the connections now seem irrelevant. "Lungwort," with its spotted leaves, reportedly cured lung diseases; "liverwort," from the shape of its leaves, helped the liver. Some were not so clearly named, although their use was clear—the brain-shaped walnut was used for injuries to the head, tongue-shaped leaves helped mouth disorders, asparagus and fennel assisted in growing hair.

For the sixteenth-century compilers of the first English herbals, books meant to identify plants and their uses, names still reflected the idea that flowers were here for our use. John Gerard, who wrote a famous herbal in 1597, believed that flowers were "for the comfort of the heart, for the driving away of sorrow and encreasing the joy of the minde." The names he gave were often descriptive and unfixed. "Herb impious" is so called because it is like "children seeking to overgrow or overtop their parents (as many wicked children do)." "Devil's bit" is named because "the Devil did bite it for envie because it is an herbe that hath so many good vertues and is so beneficent to mankind." "Cloudberry" grows where clouds are lower than mountaintops.

Recently introduced flowers from the New World sometimes carried the name of the person who had brought them, their place of origin, or even their native names. While fewer than a thousand new plants were introduced to Britain in the seventeenth century, by the end of the eighteenth century there were nearly nine thousand new introductions. The Americas proved that the number of plants existing was vast—botanists could no longer describe a few hundred of them and think they had them all, nor could the Garden of Eden, containing all the plants known to the world, be re-created in a European botanical garden, as had once been hoped. Philosophically this was tremendously important, as theorists began to acknowledge that not every plant had necessarily been created in limited quantities with a specific use for man.

The seventeenth century had seen the creation of scientific institutions and new botanical gardens. Botanists from these institutions had tried to find ways of sorting the enormous influx of plants. In the eighteenth century, the time of Carl von Linné (better known as Linnaeus), we see many new plants named after people. Descriptive names were running short, and the more detailed they were, the more cumbersome they became. Nor were medicinal virtues paramount any longer. Linnaeus proposed a revolutionary way of classifying plants with just two names: genus and species. Not all the names were given for reasons of science or respect, and Linnaeus sometimes demonstrated human weaknesses as well as strengths when he named plants. For the ambitious botanist Gronovius he named *Gronovia*, being "a climbing plant which grasps all other plants." Another name, *Monsonia*, was for Lady Ann Monson, of whom Linnaeus asked that he might "be permitted to join with you in the procreation of just one little daughter . . . a little Monsonia, through which your fame would live for ever in the Kingdom of Flora."

Nowadays we think of botanists as funny old men with magnifying glasses, but during the great age of scientific exploration they were the brightest and the best, the young, the brave, and the ambitious. All of them risked their lives and many of them died for the plants they sought. William Sherard narrowly escaped being taken for a wolf and shot while creeping after a plant. John Lawson was tortured and burned to death by Indians. Richard Cunningham was killed in Australia by aborigines. David Douglas died in a bull pit in Hawaii. George Forrest hid from Tibetan bandits for days while on the brink of death. Discomfort, illness, loneliness, and attacks from animals, insects, and hostile natives were all routine, and yet the men ventured on, because botany was the frontier of knowledge, as new as outer space is to us.

As methods of collecting became safer, and there were fewer new frontiers to explore, botany more often became the pursuit of scholars than adventurers. Nomenclature became a fussy science with its own pedantic rules, and we became more casual about the flowers we grew. It was easy to forget that someone had died for a potted plant we could pick up at our local nursery, even if we still called it by his name. Flowers became abundant and cheap—pleasing but unnecessary appendages to our more important lives. So by our success we have come full circle, and what was the unknown and the mysterious is now provided for our pleasure, as it was in the Garden of Eden.

Just after I started writing *100 Flowers and How They Got Their Names*, and within a few weeks of each other, both my older sister and my son died. My sister had always been there for me. My son, I had believed, always would be. So it was that I was tumbling through space, with the past and the future gone.

Flowers did not console me, although there were enough of them—on graves, on cards, and in sympathetic bouquets. Even the reality of their beauty, as I glanced at it and hurtled past, had no meaning. I knew with certainty it did not exist to comfort me—I was incidental to it, as I was to the universe itself.

And that, after all, is perhaps why I continued to write about flowers. Not only had their beauty not evolved for me but I suddenly realized what I had really always known. It would not make the slightest difference to them, even while I gasped at their loveliness, if I or the entire human race should die the next day. But if all the flowers died, the world we know would be no more. No flowers, no seeds, no vegetation. If they all died, we would very shortly follow. Flowers are more essential to us than we are even to one another, and if we lost them, we would lose all. Even human grief, our cries into the darkness, is nothing compared to the flowers.

If we fail to remember the history of our flowers, we know them less, and to trace their link with us is to make them part of our lives.

If we forget they are part of our lives, we may be too casual about them. The naming of flowers is no botanical game. It is the story of a relationship, a relationship of the essential to the incidental. We can call flowers what we like, we can tread on them, we can pick them. But it is always we, not they, who are incidental.

100 FLOWERS

And How They Got Their Names

ABELIA

BOTANICAL NAME: *Abelia.* **FAMILY:** *Caprifoliaceae.*

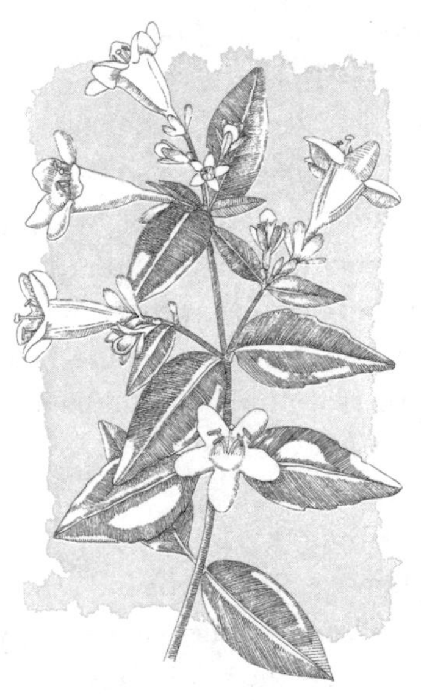

Someone should do a scholarly survey and find out if plants whose names come at the beginning of the alphabet are more often found in gardens than those that are listed farther along in the catalogs. Abelia, with its fine glossy leaves and delicate flowers, is found in most gardens. Abeliophyllum, or white forsythia, is truly a beginning plant, for it bears fragrant flowers in early spring before its own leaves, or any other, appear. Either is a good start to a garden, but although they are not related (white forsythia is a member of the olive family) both are named after Dr. Clarke Abel, who accompanied Lord Amherst on a disastrous expedition to China in 1817.

Politics, stupidity, and natural disasters were always hazards that challenged plant collectors, and Dr. Abel was hampered by them all. British access to Chinese botanical treasures was still limited to the Portuguese island of Macao and whatever plants the Chinese deigned to offer them. The British wanted to explore the interior and take back what they could find, but the Chinese understandably resented British arrogance and involvement in the opium trade. Lord Amherst was sent to negotiate an agreement with the emperor. He was, Abel said, "urged to enter the imperial presence and to prostrate" (at 6:00 A.M.), but he "declared his intention not to perform the ceremony" and the embassy was dismissed. The British asserted that they were merely refusing to "kowtow" to what Abel called "every piece of yellow rag that they might choose to consider as emblematical of his Chinese majesty," but as a result the interior of China remained closed to them until gunboat diplomacy dictated the 1842 Treaty of Nanking.

Dr. Abel collected what he could along the homeward route, but the ship, *Alceste*, was wrecked; a box of seeds and plants that had been saved was then thrown into the sea to make room for the linen of an embassy "Gentleman." What remained was captured and burned by Malaysian pirates. Abel had, however, left a few plants at Canton, and eventually the *Abelia chinensis* reached England.

Abeliophyllum, so called because its leaf (Greek, *phyllum*) is like the abelia's, has white or faintly pink flowers. The abelia has red or pink flowers from midsummer through autumn. Neither comes in any shade of yellow—perhaps luckily for the memory of a man who would not bow to that color.

AFRICAN VIOLET

COMMON NAMES: African violet, Usambara violet.
BOTANICAL NAME: *Saintpaulia*. FAMILY: *Gesneriaceae*.

There are probably more African violets in American bathrooms than in Africa. From a plant's point of view, in spite of chrome and toothpaste, warm steamy bathrooms are quite a good imitation of a tropical rain forest, and African violets flourish in them. They come from the humid forests of the Usambara Mountains in northern Tanzania. African violets grow naturally in rock crevices where small amounts of soil have been deposited and water drains away rapidly. Though they thrive on 80 percent humidity, they must not be overwatered. They get much of their water from the atmosphere through the fine hairs which cover the surface of their leaves. These hairs take in moisture from the air, like miniature roots, and also trap raindrops, separating them so the leaves don't suffocate. The roots themselves remain relatively dry.

African violets were sent to Europe in 1892, by Baron Adalbert Emil Walter Redcliffe le Tanneux von Saint Paul-Illaire, district governor of Usambara, in what was the German colony of Tanganyika. When the young governor, some say in the company of his future wife, Margarethe, was exploring his territory, he found these new plants. He collected plants or, more probably, seeds to send back to his father, Baron Ulrich von Saint Paul, a keen horticulturalist who took them to Hermann Wendland, director of the Royal Botanic Garden at Herrenhausen (Hanover). Wendland described the new plant as "of enhancing beauty . . . one of the daintiest hot house plants" and he named it *Saintpaulia*, after the two barons, father and son. He added *ionantha* because of the purple, violet-like flowers (see "Violet"). Another African violet introduced at the same time was later called *Saintpaulia confusa* because it was confused with another species!

When the British took over the colony (later known as Tanzania) after World War I, more African violets were discovered. The flowers were soon available in purples, pinks, near-reds, whites, and bicolors, with single or double flowers. There are no yellows or oranges, and the leaves vary. They can be propagated by rooting a single leaf, although some people are better at this than others. But there is no shortage of the plants in American nurseries, supermarkets, and even dime stores. Sadly though, there is a shortage of them in their native Tanzania. They can only grow in the shady rain forest, and these days forests are being felled everywhere for agricultural needs and for modern houses—with modern plumbing.

Sadly though, there is a shortage of them in their native Tanzania.

ANEMONE

BOTANICAL NAME: *Anemone*. FAMILY: *Ranunculaceae*.

Anemones used to be called "windflowers," possibly because they grew on windy sites (*anemos* is Greek for "wind"). The herbalist Nicholas Culpeper said that "the flowers never open but when the wind bloweth; Pliny is my author; if it be not so, blame him."

A more compelling derivation is from "Naamen," which is the Persian for "Adonis." Anemones were associated with Adonis, with whom Aphrodite (Venus) fell passionately in love when he was born. She tried to protect him from harm by hiding him in the underworld, but was forced by Zeus to share him with the underworld goddess, Persephone. Aphrodite was afraid he might be hurt while hunting, but of course he would not listen to her, so she could only follow him in her swan-drawn chariot. One day Adonis tracked down a huge boar and wounded it. It turned on him and gored him. Aphrodite arrived in time to hold him in her arms and weep over him as he died. Some versions of the legend say the anemone grew up from her tears and some that it sprang from his blood as it soaked into the ground, but it became the symbol of protective love that could not protect and of adventurous youth and beauty that challenged life, and lost.

Anemones were also sacred flowers, possibly the "lilies of the field" mentioned in the New Testament. Some legends say that the red petals of these wild anemones came from the blood dripping down on them from Christ's cross, and that they sprang up miraculously in Pisa's Campo Santo cemetery after a Crusader ship had brought some earth for the graves back from the Holy Land.

There were various theories about breeding them. A Dutch herbalist, Van Oosten, said that if the wind was in a southerly direction when the seeds were sown, the flowers would come out double. The "French" anemones, one story says, were stolen by a parliamentary official from the Parisian breeder who had refused to share them. The official arranged to be shown round the garden just when the anemones were going to seed. His fur-lined cloak "accidentally" slipped off his arm as he was passing the anemone bed, and his servant (previously instructed) picked it up, rolling into it some of the precious seeds.

The "Japanese" anemones were sent back to England in 1844 by Robert Fortune, who saw them growing on tombs in China and called them a "most appropriate ornament for the last resting places of the dead." These get their color from their bracts, not their petals, and they bloom in autumn, not spring. But autumn-blooming flowers are a symbol of hope and resurrection too, for gardeners believe spring is rebirth and they prepare for spring by planting bulbs in autumn. Like Aphrodite, they are consigning their hopes to the underworld, and like Aphrodite, they will hover over the fragile blossoms when they emerge. They will not always be able to protect them, but still they hope and still they believe.

ASTER

COMMON NAMES: Aster, Michaelmas daisy, Chinese aster. BOTANICAL NAMES: *Aster, Callistephus (Chinese aster)*. FAMILY: *Compositae*.

The English called European asters both "asters" and "starworts." *Aster*, Latin for "star," referred to the flower's star-like shape. "Wort" originally meant "root," and then was applied to plants that had healing properties. Asters, said the herbalist John Parkinson, were good for "the biting of a mad dogge, the greene herbe being beaten with old hogs grease, and applyed."

In 1637 John Tradescant the Younger brought North American asters back from Virginia. These do not seem to have been noticed much until they were hybridized with European starworts. They were later renamed "Michaelmas daisies" in Britain, because when the British finally adopted Gregory XIII's revised calendar, the feast of Saint Michael coincided with their flowering.

There were two botanizing John Tradescants, father and son. The elder, in 1618, traveled abroad as far as Russia. His account of the trip reveals that he had no sense of smell, and he remarks that rain leaking into the cabin had soaked and spoiled "all my clothes and beds," but his enthusiasm for flowers does not seem to have been dampened. His son, John the Younger, not only brought back the North American aster, but also collected from Barbados the *Mimosa pudica*, or sensitive plant, which, a hundred years later, may have made possible the acquisition of the annual Chinese aster. *Callistephus chinensis*, or "beautiful Chinese crown," from the Greek *kallistos* (most beautiful) and *stephanus* (a crown), is only called an aster because of its star-like flower. The Jesuit Pierre d'Incarville had been sent to China to convert the emperor, Chien Lung, to Christianity. China at the time had mostly barred Westerners, but the emperor accepted d'Incarville, who was a skilled clockmaker as well as a botanist. The priest was frustrated in his attempts to collect new plants and only got round the emperor by presenting him with two plants of the *Mimosa pudica* that he had raised from seed sent from Paris. The leaves of the *Mimosa pudica* collapse when touched and this, we are told, "greatly diverted" the emperor, who "laughed heartily." D'Incarville was now given access to the imperial gardens and was free to export plants until he died, soon afterward, in 1757.

Michaelmas was always a date of beginnings: the academic year at Oxford and Cambridge, the quarterly court session, the day for debts to be settled and annual rents (often including a goose) to be paid. In the garden both Michaelmas daisies and Chinese asters bloom in autumn, magnificent curtain calls of summer but reminders too of new beginnings after winter's sleep.

ASTILBE

COMMON NAMES: Astilbe, spirea. BOTANICAL NAME: *Astilbe*. FAMILY: *Saxifragaceae*.

The name "astilbe" probably refers to a lack of showiness in the original Chinese flowers, as it comes from the Greek *a* (without) and *stilbe* (brilliance). It is sometimes called "spirea" because it looks like *Aruncus spirea* (or *Aruncus dioicus*), commonly called "goatsbeard." Modern hybrids of red, pink, and white flowers bloom even in deep shade and are not dull at all—and neither was the life of Père Armand David, who discovered the astilbe in China.

In 1860 French and British gunboats secured a treaty from the Chinese allowing exploration of the interior and admission to Christian missionaries. Père David, a Lazarist monk, was sent to China to set up a school for a hundred boys in Peking. He was such an ardent and successful botanist that he was released from his duties so that he could collect plants. He sent thousands back to Paris, although only about one-third of his specimens survived. He cheerfully recorded his hardships in his diary: the danger of wolves obliged him to share his tent with his donkey, "though its presence there is not without inconvenience" (one wonders who got to lie down first), and the local food defied "all but the most ravenous hunger" and "must be eaten with courage," but "one man can live wherever another can."

Père David was once so ill that he was given the last sacraments, but he lived to return to Paris, where he died at age seventy-four. Other French missionary botanists were not as lucky. Père Jean André Soulié, caught between Tibetan and Chinese hostilities, was captured while packing his plant specimens and was tortured for fifteen days before being shot. Père Jean Marie Delavay caught bubonic plague and lost the use of his right arm. The plants that missionaries sent home seldom reached France or died by the time they arrived. Some of Père Delavay's boxes of plants lay unopened in a Paris museum for over fifty years.

Some of Père Delavay's boxes of plants lay unopened in a Paris museum for over fifty years.

The missionaries identified many botanical treasures that were rediscovered and introduced in the next century. Some of these were named after them. The beautiful davidia tree and the *Buddleia davidii* (see "Butterfly Bush") are called after Père David, as are Père David's deer. There is an *Iris delavayi*, and Père Soulié has primulas, a rhododendron, and a lily bearing his name. The priests' motives were not to become famous, though, or to perpetuate their own names, as some botanists have wished. These souls were driven by far different forces and, like the astilbe, their lives bloomed in the shade.

AZALEA

BOTANICAL NAME: *Rhododendron*. FAMILY: *Ericaceae*.

The difference between "azaleas" and "rhododendrons" can be as good a subject for dinner-table arguments as the difference between "hominy" and "grits"—either can amuse (or bore) the company for a whole evening, with no resolution. The name "azalea" comes from *azaleos*, Greek for "dry," and covers various species and hybrids of the *Rhododendron* genus. In fact, most azaleas do not thrive in dry ground and need to be well watered because of their shallow root system.

The first of what we now call "azaleas" to reach Europe seems to have been the *Rhododendron viscosum*, which we now call the "swamp azalea." It was sent by Reverend John Banister (see "Bluebell") to Bishop Henry Compton in London and described in 1691. In 1737 Linnaeus first applied the name to a shrub from dry habitats in Lapland, which he called *Azalea procumbens*, but which is now called *Loiseleuria procumbens* after Jean Louis Auguste Loiseleur-Deslongchamps, a physician and botanist in Paris. This first azalea, which isn't an azalea, has very small leaves and flowers and is not grown in gardens.

Meanwhile the name that was no longer applied to this shrub was applied, somewhat randomly, to some rhododendrons. On the whole, deciduous rhododendrons are often called azaleas, but evergreen "azaleas" are not necessarily called rhododendrons.

Native American azaleas are beautiful, usually deciduous, shrubs. William Bartram in his *Travels* described the "fiery Azalea, flaming on the ascending hills or wavy surface of the gliding brooks . . . that suddenly opening to view from dark shades, we are alarmed with the apprehension of the hill being set on fire. This is certainly the most gay and brilliant flowering shrub yet known." Peter Kalm, who was sent to North America to study useful plants (see "Mountain Laurel"), said of azaleas, "The people have not found that this plant may be applied to any practical use; they only gather the flowers and put them in pots because they are so beautiful."

Azaleas are some of our most used, and abused, flowering shrubs. Their natural habitat is on wooded slopes, where they will bloom through the trees with almost mystical brilliance. Indeed the Japanese believed the Kurume azalea sprang from the soil of sacred Mount Kirishina when Ninigi descended from heaven to found the Japanese Empire. We, who also have our gods, tend to plant them in parking lots of banks or supermarkets. We surround them with shredded dead bark and prune them into neat globes. There they glow like giant tonsils at the entrances of mirrored glass buildings that are lit within by fluorescent lights. We see them when we cash our checks or buy our food in plastic bags, and they are supposed to cheer us as we pass.

BABY BLUE EYES AND POACHED EGGS

COMMON NAMES: Baby blue eyes, poached eggs, fried eggs. BOTANICAL NAMES: *Nemophila, Limnanthes.* FAMILIES: *Hydrophyllaceae, Limnanthaceae.*

David Douglas was a tough Scottish explorer who botanized on the west coast of America in the 1820s. The Douglas fir is called after him. Two delicate cottage garden flowers were collected by him too. Baby blue eyes, named *Nemophila* from the Greek *nemos* (a glade) and *phileo* (to love), is a bold, celestial blue which shrinks from the open sky and scorching sun. The insouciant poached egg covers itself with hundreds of flowers which are always crawling with bees and, unless you are a bee, looks a lot like its namesake. Its botanical name comes from the Greek *limne* (a marsh) and *antheos* (a flower). Neither of these flowers can cope with Yankee summers—they come from the damp northwest coast of America and thrive in misty English summer gardens.

Douglas was a wonderful mixture of sensitivity and grit. When longed-for letters from home arrived, he was so excited that he "never slept," and got up four times in the night to reread them. He botanized in a suit of bright red Royal Stuart tartan, complete with vest, but half the time he had no proper shoes and suffered terribly from blisters. He used his gun freely to frighten everything from the Indians he encountered to the rats he caught making off with his inkwell, razor, and soap in the middle of the night, but he was accompanied everywhere by Billy, a favorite scraggy little terrier whom he adored. All his provisions had to be carried, including paper, ink, ammunition, and food. When his canoe overturned he lost everything and had to eat his plant collection. Once he wistfully noted in his journal that he had "dreamed last night of being in Regent Street, London," but a little later left his party because he felt he "must scale a peak." When he reached the top, he described the view as "beyond description striking the mind with horror blended with a sense of the wondrous." He survived snow blindness, starvation, near drowning, hostile natives, until finally, at age thirty-five, he somehow fell into a bull trap while botanizing in Hawaii and was gored to death. The little terrier was sitting by the edge of the pit and was the only witness.

Douglas had brought so many new plants to Europe that he apologized for seeming to "manufacture" them "at my pleasure." "I can die satisfied with myself," he wrote in his diary. "I have never given cause for remonstrance or pain to an individual on earth." Nowadays we look more carefully at mass importations and exchanges of plants, but in those days he, and others, died for the sake of spreading the wonders of nature wherever they could.

BALLOON FLOWER

BOTANICAL NAME: *Platycodon.*
FAMILY: *Campanulaceae.*

Vita Sackville-West described the balloon flower's puffed-up bud as "a tiny lantern, so tightly closed as though its little seams had been stitched together, with the further charm that you can pop it . . . if you are so childishly minded." Its botanical name comes from the Greek *platys* (broad) and *kodon* (bell).

The Chinese platycodon was first described by a German professor, Johann Georg Gmelin, in the court of Catherine the Great. Gmelin was sent to explore Siberia and bring his scientific and botanical discoveries back to St. Petersburg. The going was so slow and the conditions so hard that the expedition took him ten years to complete. In winter his party could hardly travel at all, and he described ice three inches thick on the windows of his cabin. In summer the mosquitoes were so bad he had to wear two pairs of gloves while writing in his journal. At Yakutsk, where he risked being captured by hostile Tartars, his cabin burned down and he lost everything, including his botanical collection.

The Japanese balloon flower, *Platycodon grandiflorus* var. *mariesii*, was discovered and named for Charles Maries, who collected for the English firm of Veitch. The *Viburnum plicatum* var. *tomentosum* is also named 'Mariesii' after him, but he is not credited with introducing many plants. He survived earthquakes, fire, and shipwreck, but his plants often didn't. He once had to replace his whole collection when a box of his seeds was in a boat that capsized and sank. Then, in China, he was robbed and his collection was again destroyed. In 1882 Joseph Hooker sent him to superintend the gardens of an Indian maharaja—where he remained until his death.

The balloon flower is one of the most coveted blue flowers for the garden. It is perennial and hardy. Those who aren't, like most gardeners, greedy for blue, can get it in pink and white as well—but that seems a waste.

• In winter Gmelin's party could hardly travel at all, and he described ice three inches thick on the windows of his cabin.

BEAR'S BREECHES

BOTANICAL NAME: *Acanthus*. FAMILY: *Acanthaceae*.

These large, spectacular thistles are widely grown in Europe but have only more recently come to American gardens. They were grown by the Greeks and the Romans, and the botanical name is from the Greek *akanthos* (thorn). The name "bear's breech" or "bear's breeches" is thought to come from the plant's soft hairy leaves or stalks, resembling respectively the rump or legs of a hairy bear. More certain is the species name *mollis*, meaning "soft," which refers to the same soft bristles.

The design of the Corinthian column is based on acanthus leaves. A story that Vitruvius tells in *De Architectura* is of an architect, Callimachus, who passed a grave on top of which a tile had been put; an acanthus plant had grown up around it, forming a circular fringe of leaves which inspired the leafy top of the Corinthian column. A sentimental Victorian version of this story says that a young girl had died a few days before her marriage and that the tile covered a basket containing the veil she would have worn. This is just the kind of tearful tale of purity that the Victorians adored, and what's wrong with a bit of sentimental elaboration? Anyway, Callimachus is credited with designing the Corinthian column, an architectural innovation because, unlike the Ionic column, which must be seen from the front, it was decorative all the way around.

The acanthus came early to Europe. Alexander Neckham, abbot of Cirencester and foster brother of Richard, Coeur de Lion, mentioned it in *De Naturis Rerum* in 1190, and John Parkinson called it a "thistle." It was among the plants collected by Joseph de Tournefort, a doctor in Louis XIV's court who botanized in Europe and the Middle East. His system of classifying plants, based on the petal structure of flowers, was used until superseded by the Linnaean system. Tournefort was quite a character. He was one of those avid gardeners (whom we all know) who does not include methods of acquiring plants within his normal moral code. Once when he climbed over a garden wall to steal plants, the irate gardener bombarded him with stones; seeming not to suffer any remorse, he merely complained that he had had to run for his life. After surviving many adventures he was run over by a carriage on the Paris street now called rue de Tournefort.

❧ The design of the Corinthian column is based on acanthus leaves.

BEAUTY BUSH

BOTANICAL NAME: *Kolkwitzia amabilis.*
FAMILY: *Caprifoliaceae.*

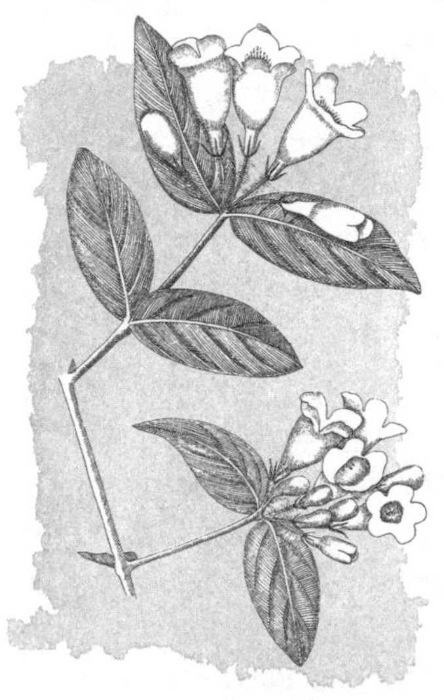

The beauty bush, although widespread, is a relative newcomer to our gardens. It was named for Richard Kolkwitz, a professor of botany in Berlin who wrote *Pflanzenphysiologie: Versuche und Beobachtungen an höheren und niederen Pflanzen einschliesslich Bakteriologie und Hydrobiologie mit Planktonkunde*, which some people may have read. "*Amabilis*" is another way of saying, in Latin, that it's lovely.

This plant has rarely been seen in the wild, which is odd because it propagates easily. Ernest Wilson sent seeds of it to Veitch's nursery in England in 1901, and it bloomed there in 1910. In 1914 Henry Veitch, who had no heirs, sold the nursery, although it had been in business for five generations, and the beauty bush disappeared too. It was introduced again to America, this time by a Dutch botanist, Frank Meyer, who found it in China between 1905 and 1918.

Meyer, many of whose expeditions were sponsored by the U.S. Department of Agriculture, was the epitome of colorful, adventurous botanists. He was a magnificent bearded figure who once walked from Holland to Italy just to see the orange groves. He nearly died in the Alps on the way. He walked all over China and, since he needed no roads, went to regions that had never been accessible to Westerners before. He sent hundreds of food plants back to America—eighteen varieties of soybean alone—and President Roosevelt used his photographs of bare, treeless hills in China to illustrate his pleas for conservation here. Meyer was a gentle Buddhist, and David Fairchild described his eyes "filling with tears" when he found that some imported bamboos had died for lack of proper care. But he was a strangely violent man too. Although he spoke eight other languages, he refused to learn Chinese, and when his interpreter was afraid to continue their journey he threw him and their coolie down stairs in a fit of rage. He died under extremely odd circumstances—his body was found in the Yangtze River. It was presumed he had fallen off a boat traveling downstream to Shanghai, but no one ever knew what had really happened, whether it was an accident, suicide, or murder.

❧ "*Amabilis*" is another way of saying, in Latin, that it's lovely.

Oddly enough he wasn't a lover of flowers. It was his fascination with plant diseases and economically useful plants that drove his long overland treks. Nevertheless, the beauty bush, with its clouds of misty pink flowers in spring, might remind gardeners of this strange, exciting man and the inaccessible places where he wandered.

BEGONIA

BOTANICAL NAME: *Begonia*. FAMILY: *Begoniaceae*.

It is possible that Michel Bégon was familiar with the flowers that bore his name, but it is not probable. Bégon was an official of Louis XIV's government in Santo Domingo and later governor of Canada, and he recommended the Minimus monk Charles Plumier to the King. Plumier named the begonia after Bégon, just as he named the lobelia, the magnolia, and the fuchsia after botanists whom he admired. He died in 1704 while waiting for a boat to take him to Peru to investigate the quinine tree's potential as a cure for malaria, the disease that killed so many of his contemporaries.

Begonias did not become important garden flowers until the nineteenth century, when South America became a rich source of new plants. Many begonias were discovered there and introduced by Richard Pearce around 1865: *Begonia pearcei* engendered today's tuberous begonias. Pearce sometimes climbed over twelve thousand feet, with no sort of equipment, to get botanical specimens.

The kind of hardships that early botanists had to undergo seem unimaginable now. They had no equipment, as we know it today, and they had to carry everything with them—this included large quantities of paper for pressing plants, and ink to make notes. They were constantly in danger and suffered from the unremitting attacks of insects. Again and again we read of explorers being plagued with mosquitoes, gnats, or fleas. Why they did not carry protective gauze or netting, which had been available since classical times, is hard to understand. Indeed the Greek historian Herodotus tells of Egyptian fishermen wrapping themselves in their nets as a protection from mosquitoes, and Cleopatra is described as stretching her mosquito net on the Tarpeian Rock before getting down to the business of seduction.

> ❧ Pearce sometimes climbed over twelve thousand feet, with no sort of equipment, to get botanical specimens.

Pearce died of mosquito-borne yellow fever in Panama at the age of thirty. Soon after Plumier's time quinine was used to cure malaria, but many botanists died anyway because the connection between mosquitoes and disease was not deduced until the beginning of the present century.

BLEEDING HEART

COMMON NAMES: Bleeding heart, lady in the bath. BOTANICAL NAME: *Dicentra spectabilis.* FAMILY: *Fumariaceae.*

Bleeding heart does look like a dripping heart. But if you turn the flower upside down and pull it slightly open, it also looks exactly like a "lady in the bath," which it is sometimes called. Its botanical name is from the Greek *di* (two) and *kentron* (spur). *"Spectabilis"* is saying, in Latin, that it is spectacular, which it is.

Robert Fortune introduced bleeding heart once the Treaty of Nanking, in 1842, gave plant collectors some access to China. Fortune set out to explore an unknown world, equipped with a Chinese dictionary, a stick loaded with lead that he called a "life preserver," and three of the new "Wardian" glass cases, which had recently been invented by Nathaniel Ward. He succeeded in sending back many of our greatest garden treasures.

The carrying cases were invented accidentally when Ward buried a chrysalis in a closed bottle containing earth. He had intended only to watch the moth develop, but he noticed that a small fern grew and prospered in the bottle, and he hit upon the idea of using similar airtight containers to protect plants from salt spray, lack of water, and changes in temperature on the long sea voyages home. It completely revolutionized the transportation of live plants from abroad.

• Fortune went into the still-forbidden interior of China to smuggle out plants, dressed in Chinese clothing complete with false pigtail.

Fortune went into the still-forbidden interior of China to smuggle out plants, dressed in Chinese clothing complete with false pigtail. His adventures included being set upon by angry crowds and robbed, falling into a wild boar trap, and being attacked by pirates, whom he held at bay while the crew of the junk hid below. When trespassing into the interior, he did not dare eat at inns because his lack of skill with chopsticks would have betrayed him. But although he collected many treasures, he was not flattering about the Chinese. They were, he said, filled with "the most conceited notions of their own importance and power; and fancy that no people, however civilised, and no country, however powerful, are for one moment to be compared with them." This did not prevent him from collecting some of their loveliest garden treasures, most of which were the result of centuries of Chinese breeding and cultivation. He did not, like later explorers, go deep into the interior to collect truly wild plants, but we have to thank him for the many flowers he brought from what he called the "central flowery land."

BLUEBELL

COMMON NAMES: Virginia bluebell, English bluebell, Spanish hyacinth. BOTANICAL NAMES: *Mertensia virginica, Endymion non-scripta, Hyacinthoides non-scripta.* FAMILIES: *Boraginaceae, Liliaceae.*

Several flowers, including Canterbury bells and harebells, are sometimes called "bluebells," but the two that concern us here are the American bluebell and the English bluebell. Both, in their respective countries, turn spring woodlands into shimmering sheets of blue, as if the sky itself were reflected back onto the earth. They are not related, except in their capacity to perform this miracle.

The American bluebell is called *Mertensia virginica* after Franz Karl Mertens, a German botanist and director of a business school in Bremen. The name is sometimes wrongly attributed to his son, Karl Heinrich, for he too was a botanist, though neither has much to do with the damp Virginia woodlands where the mertensia flourishes.

The Virginia bluebell was first sent back to Europe by John Banister, a young clergyman sent to Virginia by Bishop Henry Compton to be in charge of the spiritual health of the American colonies. Although he specialized in freshwater snails, Banister was also a botanist who wrote about a world of plants so strange that he feared others might think them "chameras" of his own mind. He died while botanizing, apparently shot by a soldier who mistook him for a wild animal.

William Turner called the native English bluebell "commune Hyacinthus" or "crowtoes" and said that its roots made very good glue. John Gerard called it "Hyacinthus Anglicus" and recommended the roots as starch for stiffening ruffs. It does not have the "inscription" *AI AI* on the petals (see "Hyacinth") so it was called *non-scripta*. Another of its names was "scilla," from the Greek for "sea squill." The Spanish scilla, which has flowers all round the stem, is sometimes confused with the English bluebell, which has flowers only along the lower side of its drooping stalk, because both are found wild and sometimes hybridized in English woods.

❧ William Turner called the native English bluebell "commune Hyacinthus" or "crowtoes" and said that its roots made very good glue.

The name *Endymion* comes from the Greek youth Endymion, with whom Selene, the moon, fell in love. He was a shepherd, and she looked down on him in the fields and was smitten by his extraordinary beauty. She managed (by her own or Zeus's exertions) to make him sleep forever, so she could always flicker over his body and kiss him where he lay. Actually, bluebells, massed in a wood, bring the whole noontime sky flickering down on them, not just the moon, and, like Endymion, they will do it forever if they are left alone.

BOUGAINVILLEA

BOTANICAL NAME: *Bougainvillea.*
FAMILY: *Nyctaginaceae.*

Bougainvillea was named for Louis Antoine de Bougainville, who was commissioned by Louis XV to circumnavigate the world and obtain any unclaimed territory to compensate for French losses to the British in North America. But it was the botanist Philibert Commerson who actually discovered and named the bougainvillea. Commerson accepted Bougainville's invitation to accompany him around the world to divert himself from his despair when his wife died in childbirth.

The spectacular new vine that Commerson called after his friend and captain originates in South America, but Commerson saw it in Tahiti ornamenting houses. The two men were captivated by Tahiti and the inhabitants' spontaneous and public love-making. Commerson noted that their "chief god" was love, but that their religion included an occasional human sacrifice. The Tahitians found the Europeans intriguing too, particularly Commerson's cat, for which they offered to exchange their loveliest maiden. Commerson, some would think to his credit and others foolishly, refused.

A fellow voyager saw Commerson spitting blood as he worked at night, and he was seasick much of the time, complaining that his meals were "loans." But at every stop he botanized, accompanied by his young assistant, Jean Baret, with whom he collected over three thousand new plants. On Tahiti, Baret attracted and was seized by a local chieftain who may have had a particularly discerning eye. For in the struggle that ensued, Baret was revealed to be not a young man but a young woman (see "Hydrangea"). Some accounts say that she had deceived Commerson, others that she had persuaded him to take her along in disguise because she so longed to go around the world. In some accounts her name and the name of Commerson's housekeeper in France seem to be the same.

In the struggle that ensued, Baret was revealed to be not a young man but a young woman.

Bougainvillea has a generosity of bloom that comes from southern seas. Where the climate suits it, it spreads curtains of pink, purple, and white wherever it can reach, forming welcoming banners however bare the wall it climbs, however hot and brown the earth. The color comes from its bracts, which are leaf-like organs that look like flower petals; the flower itself is yellow and quite insignificant. The colored curtains endure so long they seem like permanent backdrops to whatever scene they decorate, from Tahitian love feasts to modern terraces. They look wonderful with a girl and a cat or two sitting in front of them.

BUTTERFLY BUSH

COMMON NAMES: Butterfly bush, summer lilac.
BOTANICAL NAME: *Buddleia*. FAMILY: *Loganiaceae*.

The name "buddleia" is after the Rev. Adam Buddle, a rector in Essex, England. There was a long tradition in England associating botanists and gardens with the clergy. Gilbert White, Canon H. N. Ellacombe, Charles Kingsley, and William Wilks are only a few of the better known horticultural clerics. Clergymen were often isolated in small villages, leading quiet, leisurely lives, and could satisfy their intellectual curiosity, as well as use their classical educations, with botanical research. They believed that to study plants would bring them closer to understanding God's universe, and the innocence of Eden. As Charles Kingsley, who wrote *The Water Babies*, affirmed, "All natural objects . . . all forms, colours and scents . . . are types of some spiritual truth or existence."

In 1708, Buddle wrote an *Herbarium* of British plants, supporting the botanical systems of John Ray and Joseph de Tournefort. He was an authority on mosses, but that did not deter Linnaeus from giving his name to a shrub, *Buddleia globosa*, which was introduced from Peru in 1774. It isn't tough enough to survive New England winters, but its globular golden flowers are very attractive and it is still found in older English gardens. The hardy buddleias, introduced later from Asia, are widely grown in Britain and North America.

The most popular buddleia, *Buddleia davidii*, was called after Père Armand David, a Jesuit missionary who explored in China (see "Astilbe"), though it was actually discovered by Père Jean André Soulié. It was sent to Kew in 1887 by Dr. Augustine Henry, an Irish customs officer in Shanghai and the assistant medical officer at Ichang. He was, in advance of his time, worried about air purity and deforestation, and described the Chinese hillsides, denuded of trees, "for all the world like a nightmare dream of telegraph poles gone mad and having a mass meeting." When he returned home he became professor of forestry at Dublin until his death in 1930.

They attract butterflies, sometimes so successfully that the bush looks as if it were flowering with butterflies.

The buddleias with their lilac-like flowers are particularly popular these days because they attract butterflies, sometimes so successfully that the bush looks as if it were flowering with butterflies, attached by their heads, crazily drinking the sweet nectar, their petally wings fluttering from the branches. It's a wonderful sight and surely we must, like Adam Buddle, be reminded by it of the infinite mystery of the universe.

CALIFORNIA POPPY

BOTANICAL NAME: *Eschscholzia.*
FAMILY: *Papaveraceae.*

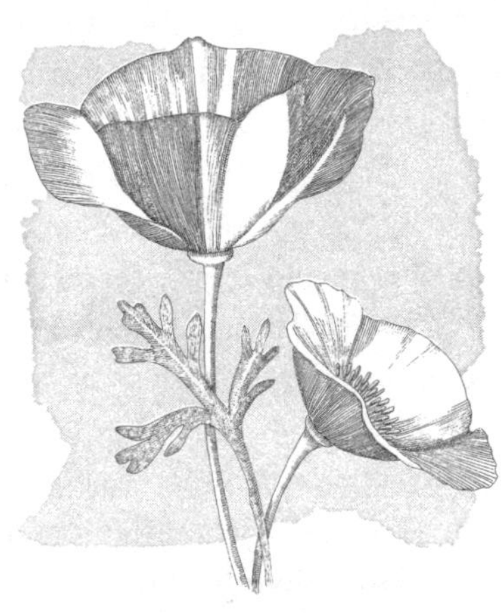

When Spanish explorers saw the hills of California blazing with flowers they called the country the land of the "Golden West," not because of the wealth that was later sought there, but because of the golden poppies. As we all know, the other western gold, and another member of the poppy family (see "Poppy"), have caused untold human misery. But the California poppy was called by the Spaniards "*copa de ora*," or cup of gold, and the evils of gold or of poppies had no connection with it.

The California poppy was taken back to Russia by Adelbert Chamisso, the botanist on a voyage led by Otto von Kotzebue (who circumnavigated the globe three times). Chamisso named the poppy after a Prussian doctor on the ship whose name was Dr. Elsholz, which was Russianized to Eschscholz or Eschscholtz. Western botanists called his namesake *Eschscholzia*, or sometimes *Eschscholtzia*.

This expedition around the world, aboard the Russian ship *Rurik*, was organized in 1815, after the end of the Napoleonic Wars. Chamisso was a French revolutionary refugee who had grown up in

Berlin. In 1814, just before the voyage, he published an allegory about a man who sold his shadow to the Devil in exchange for a limitless purse, and thereafter wandered endlessly searching for peace of mind. Chamisso himself was a wanderer by nature who joined the three-year expedition with all its hardships and kept a detailed journal of the trip that vividly conveyed its frustrations and joys, which included Sunday concerts at which the ship's Bengali cook played the violin.

They reached the Bering Strait, searching for the elusive Northwest Passage and anchoring at what are now the ports of Chamiss and Kotzebue. They went on to explore the Pacific islands, one of which the captain named Eschscholtz Atoll. This name was retained until 1946, when the island became well known as Bikini and acquired associations with the sadder side of human history.

The California poppies are hardy but look fragile, as fragile as the good dreams of those who wandered the earth in search of knowledge. It is the kind of poppy that does no harm and the kind of gold that evokes no greed. To plant them in our gardens is to add beauty to the world without the slightest effort. One simply sprinkles the seeds on the ground, and the reward is a carpet of gold.

❧ The California poppy was called by the Spaniards *"copa de ora,"* or cup of gold.

CAMELLIA

BOTANICAL NAME: *Camellia*. FAMILY: *Theaceae*.

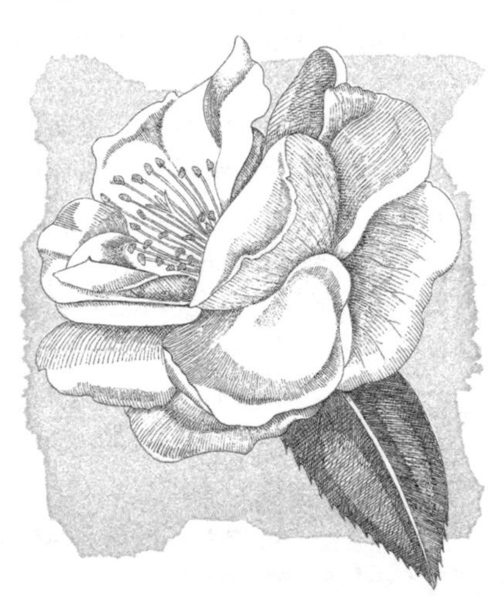

The camellia is not, as it is often supposed, called after the Lady of the Camellias, a famous nineteenth-century French courtesan. She was a country girl who came to Paris to make her fortune, aided by her beauty. Her name was Madeleine du Plessis, and her beauty and her early death inspired a novel, a play, and an opera about love, social codes, and purity of heart. She inspired great art, but she was a courtesan, and in her story there is a vein of earthiness combined with the idealism of human love. She always carried a bouquet of camellias that were on twenty-five days of the month, white, and on the other five days, red. Was that because, on five days out of the month, a woman might be what the Victorians called "indisposed"?

Madeleine du Plessis was the model for Alexandre Dumas's heroine Marguerite Gauthier. The camellia was enhanced by human art too, and is the result of centuries of cultivation in China, where this relative of the tea plant was a garden treasure long before it was ever brought to the West.

Georg Josef Kamel, for whom the flower was actually named, has not really had the benefit of the honor. His association with the flower is hardly remembered, and he probably never saw a camellia anyway. Linnaeus named the flower for him, changing the *K* in his name to a *C* to fit the Latin alphabet (which has no *K*). Kamel was a Moravian Jesuit missionary who studied the plants and animals of the Philippines. Linnaeus first called the camellia *Thea sinensis*, or "Chinese tea," but in the second volume of his *Species Plantarum* he changed it to *Camellia japonica*. If the two volumes had not been considered the same work by botanists, *Thea* would have been its correct name and Kamel would not have been remembered at all.

The first camellia was sent to England by James Cunningham, who was the only survivor of a massacre of East India Company officials in 1705. Lord Petre, whom Peter Collinson had called "the best botanist in England," killed it in too hot a greenhouse. But his gardener, James Gordon, had taken cuttings, which survived.

Camellias arrived in America in the late eighteenth century and soon became so much a part of Southern gardens they seem to be native there. Maybe their perfect purity is just a little bit deceptive, because they do not have a sweet scent, and no one has been able to breed one into them. They are flawless flowers without the mystery of perfume, and indeed that is why Marguerite Gauthier preferred them, for scented flowers made her cough. Like herself, they are utterly beautiful, but far from perfect.

CANDYTUFT

COMMON NAMES: Candytuft, candy mustard.
BOTANICAL NAME: *Iberis umbellata.* FAMILY: *Cruciferae.*

One would think candytuft got its name from its pink and white flowers, which look like confectioners' sugar lollipops. Actually the name comes from Candia, or Crete, from where it was imported to England in Elizabethan times.

Lord Edward Zouche was credited with bringing it to England. Lord Zouche was said to have spent so much on his garden that he became impoverished and had to travel abroad to "live cheaply." For a while the famous botanist Matthias de l'Obel was employed by him as his gardener in Hackney before he became gardener to King James (see "Lobelia"). Zouche was on the New England and Virginia councils, was lord warden of the Cinque Ports, and was a friend of the poets Ben Jonson and William Browne. None of this tells us much about him, but what does was that he was the only peer to acquit Mary, Queen of Scots, at her trial and to dissent from the death sentence for her. He brought seeds home to John Gerard "for which," Gerard notes, "I think myself much bounde unto his good Lordship." Among them were seeds of the candytuft or, as it was then called, *Thlaspi candiae* (or "Cretan cress").

John Parkinson called it "Treacle Mustard," and it was used as a cheap condiment. It grew "in Spaine and Candie, not farre from the Sea side." Its botanical name, *Iberis umbellata*, means it has flowers in umbels, or tufts, and comes from Spain (Iberia).

The annual common candytuft is very easy to grow and consequently has always been a standby of children's and cottage gardens. It doesn't like hot weather, so it isn't seen in eastern American gardens much. The perennial candytuft, or *Iberis sempervirens* (Latin *semper*, "always"; *virens*, "alive") is more popular because in spring it forms nice evergreen mats covered with tiny white flowers.

Candytuft is tough, but isn't a particularly interesting flower, apart from its name. Though uninteresting inhabitants of the world can serve a useful purpose, horticultural show-offs probably won't plant it—they have their own ways of enjoying themselves.

• Lord Zouche was said to have spent so much on his garden that he became impoverished and had to travel abroad to "live cheaply."

CARNATION, PINK, SWEET WILLIAM

BOTANICAL NAME: *Dianthus.*
FAMILY: *Caryophyllaceae.*

Dianthuses are ancient flowers, and derivatives of their different names and forms are various. The Greek botanist Theophrastus, who first classified plants according to their form and structure, called them *"dianthus,"* from the Greek *dios*, "divine," and *anthos*, "flower." The most common garden dianthuses are carnations, pinks, and sweet Williams.

Some scholars think that the name "carnation" is from *coronation* or *corone* (flower garlands), as it was one of the flowers used to make ceremonial crowns in Greece (see "Spirea"). Others say this name comes from *carnis* (flesh), referring to the color of the flowers, or possibly from *incarnacyon* (incarnation), referring to the incarnation of God, made flesh. The flowers were also symbolic of marital bliss and fecundity, and at his wedding ceremony Maximilian of Austria was instructed by the bishop of Trèves to search under his bride's wedding dress for a carnation hidden there—which he did, we are told, first tentatively, and then with increasing enthusiasm.

Pinks first came to Britain in the middle of the sixteenth century. It seems that they would have been so named because their color is pink, but actually it was the reverse. Pink was not a specific color until the eighteenth century, and almost certainly came from the name of the flower. Some say the word comes from the Middle English *poinken*, which originally meant "to pierce holes" in leather or cloth, and then came to mean decorating the edges—in a similar manner to the pinked edges of dianthus petals.

The name "clove pink" or "clove gillyflower" is probably derived from the French *clou de girofle*, or "nail of the clove tree," once called *Caryophyllus* (from the Greek *caryon*, "nut," and *phyllus*, "leaf"). In any case the pink's clove-like fragrance led to its association with Crucifixion nails, because cloves are shaped like nails. Sometimes the infant Jesus is shown in paintings innocently playing with a carnation or pink, as a dreadful reminder of his future. This could also, of course, refer to the fact that he was "God incarnate."

The name came from the French *oeillet*, "eye," which became "Willy."

Sweet Williams, the biennial pinks, are also a mystery of nomenclature. Some say the name came from the French *oeillet*, "eye," which became "Willy" and then "William." Some say it was from Saint William, whose festival is on the twenty-fifth of June, when the flowers bloom. John Gerard suggested they might have been named for his contemporary William Shakespeare, whom he generously called "the greatest man" of the time. But maybe Gerard is an example of his own belief that he "that doth looke upon beautifull things [cannot] have his minde not faire."

CHRISTMAS ROSE

BOTANICAL NAME: *Helleborus.*
FAMILY: *Ranunculaceae.*

The Christmas rose blooms at Christmastime. It really does, and even in Pennsylvania the flowers push out of the snow. The blooms last for weeks and the plant lasts for years.

It is supposed to have bloomed outside the stable at Bethlehem, although scholars have taken pains to discover that it is not native to the Holy Land. While applauding their industry, some people don't care anyway, and still think of the stable in the snow, the hovering angels, the kneeling donkey, and the other details that may not fit climate, gravity, or animal behavior patterns. The Christmas rose fits nicely into the story, for its legend tells us that a little country girl visited the stable and wept because she had nothing to give the Christ child. Her tears fell in the snow and a hovering angel landed and showed her the Christmas rose poking through the snow to use as her gift.

It's actually not at all the thing to give to a newborn baby as it's very poisonous, and its botanical name is from the Greek *hellein* (to kill) and *bora* (food). It was used from ancient times (with caution) as a medicine, especially to cure worms in children. Gilbert White mentions it in his letters from Selborne but warns that it is a "violent remedy" that kills the worms but might also kill the patient. John Gerard said it was good for "mad and furious men . . . and for all those that are troubled with blacke choler, and molested with melancholy."

It was used in ancient Greece by Melampus to cure the daughters of Proetus, king of Tiryns. These young women had treated with contempt a statue (some say of Hera, others of Dionysus). As a punishment they were deprived of their senses and streaked naked through the Peloponnesus. Melampus, who was a shepherd, somehow got them to stop long enough to drink milk from his goats, which had eaten hellebore, and they were cured. Melampus asked for, and got, quite a bit of Proetus's kingdom for curing his daughters.

The Christmas rose is surely a miraculous plant, regardless of its name's unscientific origins. For one thing, its seeds are spread by, of all things, snails. They eat the oil covering the seed and carry the rest away in their slime. Certainly a different process than that of being born from tears, but slime and tears glitter equally on moonlit nights and both are mysterious. It's certainly no normal plant, as anyone who has come out on a January morning and looked at it will attest. The flowers are literally frozen solid and yet, when the ice falls away, the petals are soft and fresh as spring blossoms. There is surely a scientific explanation for this, but some just marvel at it anyway.

CHRYSANTHEMUM

COMMON NAMES: Chrysanthemum, mum, tansy.
BOTANICAL NAMES: *Chrysanthemum, Dendranthema.*
FAMILY: *Compositae.*

By chrysanthemums we usually mean mums, the popular fall-blooming perennial or hothouse plants originating in China. Actually several flowers we call daisies, such as the ox-eye daisy, the painted daisy, and the Shasta daisy, are technically chrysanthemums (see "Daisy").

The name "chrysanthemum" comes from the Greek *chrysos* (gold) and *anthos* (flower). The Mediterranean *Chrysanthemum coronarium*, from the Latin *coronarius* (used for garlands), was a golden-yellow flower from which garlands were made to protect against demons. It was also called *Dios ophrya* (God's eyebrow). The European feverfew or featherfew (the medieval *Tanacetum*, or "tansie," now *Chrysanthemum parthenium*) was widely used as an antipyretic.

The Chinese chrysanthemum, originally a daisy-like wild plant, had been cultivated in Chinese gardens for more than twenty-five hundred years before it came to the West. The fourth-century poet T'ao Yuan-Ming had a famous chrysanthemum garden to which he retired after refusing a high government post. He preferred to "pick chrysanthemums from the hedges," entertain his friends, and get drunk.

Chrysanthemums symbolized a scholar in retirement, though not necessarily a recluse. Infusions of the petals and leaves made wine and medicine, and the dew collected from them was supposed to promote longevity. They were considered one of the four "noble plants" (the other three being bamboo, plum, and orchid). About A.D. 400, Zen Buddhist monks took chrysanthemums to Japan, where they eventually became the symbol of the Mikado, represented by an insignia that looked like the Rising Sun but was in fact a sixteen-petaled chrysanthemum.

❧ It was also called *Dios ophrya* (God's eyebrow).

The first "garden" chrysanthemum (*C.* × *morifolium*) was exhibited in England in 1795. In the nineteenth century, John Reeves, tea inspector for the East India Tea Company in China, sent home chrysanthemums and botanical drawings by Chinese artists. Robert Fortune sent home the Chusan daisy, which became the pompom chrysanthemum, so called because in France, where it was first grown, it looked like pompoms on sailors' hats.

Since chrysanthemums are short-day flowers, they are well adapted to greenhouse cultivation and can be hoodwinked into blooming at any time of year by decreasing the amount of light they receive. Outdoors they bloom in autumn. One ardent chrysanthemum grower is said to have sued his township because a streetlight shining on the flowers at night prevented them from blooming. They are used freely in houses in England and America, but in Italy, perhaps because of the time of year they bloom, they are associated with the dead and are unacceptable in any other context.

CLEMATIS

COMMON NAMES: Clematis, virgin's bower, old man's beard. BOTANICAL NAME: *Clematis*. FAMILY: *Ranunculaceae*.

Clematis vines were growing all over the world, both wild and in gardens, before Linnaeus conclusively named them (from the Greek *klema*, "a twig"). The Swiss botanist Kaspar Bauhin, in a compendium of the six thousand plants then known, had listed a "clematitis," and John Parkinson changed the spelling to "clematis," describing its seeds as a "round feather topt ball." The Japanese called them "wire lotus." John Gerard called them "Travellers-Joy" and said the plants had no use but were "esteemed onely for pleasure, by reason of the goodly shadow which they make with their thicke bushing and clyming, as also for the beauty of the floures."

The first Asian clematis to reach Britain seems to have been the *Clematis florida*, which grew in the garden of the famous Quaker, Dr. John Fothergill. Fothergill humbly described his garden as "a paradise of plants of small extent, whose master . . . has at least a burning love of botany." But Sir Joseph Banks said that "no other garden in Europe, royal or of a subject, had nearly so many scarce and valuable plants."

Fothergill's clematis was cultivated in Japan but was a native of China, where most of the large-flowered clematis varieties originated. The enormous-flowered Chinese *Clematis lanuginosa* (from the Latin lanuginosus, "woolly") was imported by Robert Fortune (see "Bleeding Heart") in 1850.

The most popular clematis grown is the gorgeous purple *C.* × *jackmanii*. It was bred in the Jackman nursery in 1858 and is generally believed to be a cross between three other varieties. George Jackman published *The Clematis as a Garden Flower*, in which he suggested planting a clematis garden with the vines trained over picturesque old tree stumps. By then though, a new fashion had started of pegging down clematis vines to cover the ground and fill flower beds. William Robinson also suggested they should be allowed to grow through shrubs such as azaleas, "throwing veils over the bushes here and there."

The new British hybrids were introduced to America in the 1890s, but the British "wild" garden style of Gertrude Jekyll and William Robinson never really became fashionable here, probably because America was wild enough as it was. Andrew Jackson Downing, the American landscape gardener, said that clematis "are capable of adding to the interest of the pleasure ground, when they are planted so as to support themselves on the branches of trees." They do not seem to have been allowed to sprawl over the flower beds.

Clematis are most often seen nowadays growing up mailboxes, where they hang nicely in "veils." The flowers are breathtakingly beautiful, especially when seen up close—which we have an opportunity to do whenever we collect our junk mail and bills.

COLUMBINE

COMMON NAMES: Columbine, granny's bonnet.
BOTANICAL NAME: *Aquilegia*. FAMILY: *Ranunculaceae*.

At first it seems odd that the common and favorite name of the flower of cuckoldry and the flower of the mysterious doctrine of the holy dove should be "granny's bonnet," even though it is shaped like an old-fashioned bonnet. Columbines remind us of bonneted little old women, nodding and gossiping, huddled under walls and in corners, where these flowers love to grow. But maybe the name is not so odd after all. For old women, bobbing and trembling, have memories under those bonnets: they might well have known the excitement or agony of cuckoldry (depending on the part they played); they may have been mothers who saw the lives of their sons given to mysterious idealism (their own or someone else's); they may have buried babies, and hoped for rebirth. The columbine's bonnets, maybe, could represent it all.

Its botanical name is *Aquilegia*, either from the Latin *aquila*, "eagle," because the spurs were like long eagle's talons, or, more probably, from *aquilegus*, "a water collector" or "water container." Greek jars for holding liquids and oil were often pointed at the base and buried upright in the ground to keep the contents cool, and they did look a lot like the spurs of the columbine flower. Spurs, like all horns, also symbolized cuckoldry and disloyalty.

The name "columbine" comes from the Latin *columba*, "dove." Held upside down the flower looks a bit like a ring of doves drinking. John Gerard said the leaves too were "the shape of little birds." The flower, with one petal and sepals removed, also resembles a hovering dove.

In old paintings and tapestries, columbine represented the dove of peace, symbol of the Holy Spirit, and it can be seen in religious works poking up through grassy foregrounds or the interstices of terraces. Many garden flowers had symbolic religious meanings; paintings depicted them blooming and fruiting together in a perpetual spring that knew no other seasons, in a Garden of Eden that was thought to have existed somewhere in reality. Enclosed medieval gardens of Paradise were attempts to literally re-create a paradise of all the plants in the world (and until Darwin's time these were thought to have been the same since the day of Creation).

In the medieval paintings, animals lie peacefully together on grassy swards, surrounded by flowers, including columbines. Indeed the columbine was also called the "herbe wherein the Lion doth delight" because it was believed that lions, at least lions in Paradise, liked to eat it. Whether, when they tried, it fluttered out of their reach, like doves, we do not know. At any rate, for a bit they were all there in Eden together.

CRAPE MYRTLE

COMMON NAMES: Crape *or* crêpe myrtle, China berry. BOTANICAL NAME: *Lagerstroemia*. FAMILY: *Lythraceae*.

Although crape myrtle grows all over the American South, it was introduced, probably from a Chinese cargo ship, by a Frenchman, André Michaux. It was named by Linnaeus after Magnus von Lagerström, generous benefactor to Uppsala University, where Linnaeus taught. Lagerström once brought Linnaeus a rhinoceros-horn cup from China carved with fruit, flowers, and lizards. This cup was among the possessions and papers of Linnaeus bought by Sir James Smith after the botanist's death. When the Swedes realized that all of Linnaeus's effects would be taken to England, there was a great stir—one story says that the Swedish navy chased the boat carrying them to England. This is probably not true, but a nice engraving exists which shows the English ship out-sailing its Swedish pursuers. In 1970 the Linnaean Society returned the rhinoceros-horn cup (but not the papers) to Sweden in honor of the king's eighty-eighth birthday.

Michaux was a Frenchman who took up botany after his beloved wife died. His adventures in Persia included being robbed of all his possessions twice and left without his shoes, making it hard for him, he said, to botanize on the hot sand. In 1785 he was sent by the French government to America to collect American plants, particularly trees that could help in reforestation. With his fifteen-year-old son François he set up nursery gardens in New York and Charleston to house his American collections and introductions from abroad, including the Lombardy poplar and the ginkgo. Michaux stayed ten years, exploring from Florida to Canada. A water primrose and an oak are named *michauxiana* and *michauxii* in honor of father and son. Michaux sent thousands of trees back to Versailles, though many did not survive the voyage home. He himself almost died in 1796 when he was returning to France and his ship was wrecked off the Holland coast. Michaux was washed ashore, unconscious, lashed to a spar. He survived, along with his collection of dried plants. After all his years of service, the revolutionary government refused to pay the salary he was owed, and he finally died of fever in Madagascar.

❧ Michaux was washed ashore, unconscious, lashed to a spar. He survived, along with his collection of dried plants.

Crape myrtles were sometimes called "China berries," and the berries were used for rosaries. They are called "crape" or "crêpe" myrtles from the Latin *crispa* (curled), as the blossoms are crinkled, like crêpe paper or crêpe de Chine. They were often planted near stables as they were supposed to keep away flies.

CROCUS

BOTANICAL NAME: *Crocus*. FAMILY: *Iridaceae*.

Crocuses flower about Valentine's Day, just when we need a reminder that winter is over and we really do love one another after all. "Krokos" was the Greek name for the autumn-flowering saffron crocus, which has been cultivated from antiquity, but which hardly anyone grows today. The nicest legend about its origin is of Zeus and Hera making love so passionately that the heat of their ardor made the bank on which they lay burst open with crocuses.

The first spring crocuses were sent to England from France by Jean Robin, curator of the Jardin du Roi in Paris. John Gerard's famous *Herball* describes the "wilde or Spring Saffron" as a novelty compared to the "best-knowne" saffron. Saffron crocuses are pale purple, and Gerard talks about the new colors of white and a "perfect shining yellow colour, seeming a far off to be a hot glowing cole of fire." The purple spring crocus, he says, "lovers of plants have gotten into their gardens" too.

Saffron was always a valuable crop. Measured ounce for ounce it was often more valuable than gold; it takes four thousand stigmas to make just one ounce of saffron. In the Middle Ages it was sometimes used instead of real gold leaf to illuminate missals. The rich used it for flavoring food (the poor had to make do with calendula petals), and it was also thought to be "good for the head." Apparently this had its own dangers: Joseph de Tournefort "saw a lady of Trent . . . almost shaken to pieces with laughing immoderately for a space of three Hours, which was occasioned by her taking too much Saffron."

The saffron crocus's name originally comes from the Arabic *za'faran*, used in the Middle East from ancient times. The best story about its introduction is Hakluyt's, in *English Voiages* (1589). He says "a pilgrim, proposing to do good to his countrey, stole a head of Saffron, and hid the same in his Palmer's staffe, which he had made hollow before of purpose, and so he brought the root into this realme with venture of his life, for if he had bene taken, by the law of the countrey from whence it came, he had died for the fact." It was dangerous stuff to fool with: conviction for adulterating saffron carried the death penalty, and in the fifteenth century Jobast Findeker was burned alive in Germany, along with his bags of impure saffron.

Crocus roots, or corms, are actually thickened stalks, and these were brought over to America by settlers. Mice and rats love them, but a few must have arrived safely and come up in cheerful clusters around cabin doors after the first grueling winters. Squirrels dig them up too, and birds love to peck the petals off—although they are, like Gerard, fonder of the yellow ones.

CYCLAMEN

COMMON NAMES: Cyclamen, sowbread.
BOTANICAL NAME: *Cyclamen.* FAMILY: *Primulaceae.*

The way cyclamen got its name is perhaps less colorful than debate over its use. Pliny the Elder's famous first-century *Natural History* claimed that the roots of a plant he called *aristolochia* were used by fishermen to poison fishes. The Renaissance botanist Nicolo Leoniceno disputed this, saying in 1492 that the root Pliny meant was a wild European cyclamen, which he himself had seen used by fishermen in Campania. Pandolfo Collenuccio, in a defense of Pliny, argued that fishermen used several plants (including cyclamen) this way and there was no reason to suppose Pliny mistaken.

Before the science of botany developed, this kind of theoretical debate preoccupied scholars, and cyclamen's medicinal uses were tailored to fit many theories. During the Renaissance, the influential Doctrine of Signatures, popularized by Paracelsus, held that the appearance of different plants conveniently indicated the use for which they had been created. Cyclamen, because it had a leaf shaped much like an ear, was used to treat earaches.

The English botanist William Turner warned that cyclamen was such a potent aid to childbirth that it was dangerous for pregnant women even to step over cyclamen roots. Poor Turner knew about childbirth. He complained that his living quarters were so crowded that "I can not go to my booke for ye crying of childer and noyse yt is made in my chamber." He wrote the first popular English herbal, published in 1551, which called cyclamen "Sawesbread."

The name "sowbread" refers to the supposed use of cyclamen tubers as food for pigs, although Canon H. N. Ellacombe, in 1895, said that some pigs had got into his garden and dug up a bed of cyclamens without eating any of them. The name "cyclamen" comes from the Greek *kyklo* (circle) and probably refers to the seed stalks, or pedicels, which after flowering curl up and ripen among the leaves. The Greek name for cyclamen was *chelonion* (tortoise) because the tubers look like little turtles.

The wild European cyclamens are enchantingly diminutive versions of the gross hothouse cyclamens more often grown today. These are descendants of the Persian cyclamen, which came to Britain in the 1650s, and are an example of freak gigantism that Victorian plant breeders were able to exploit. They make handsome houseplants that the Victorian writer John Loudon claimed lived for years and were "easily raised from seed" to produce "from fifty to eighty blossoms." They do not have the magic of the meek (but hardy) wild cyclamens that will grow in America south of climatic zone 5 and that will, if content, spread to make brilliant clumps. Most of us feel triumphant if we manage to keep the conservatory cyclamens alive at all for more than one season, let alone growing them from seed. The Victorians must have been better gardeners, or better liars, than we are.

DAFFODIL

COMMON NAMES: Daffodil, narcissus, jonquil.
BOTANICAL NAME: *Narcissus.* FAMILY: *Amaryllidaceae.*

The difference in meaning between the names of daffodils, narcissi, and jonquils is still unclear, but we seem to agree that all daffodils are narcissi, though not all narcissi are daffodils, and it has to do with length of trumpet and number of flowers per stem.

The confusion over the name "daffodil" may have started early, when the British, who preferred the imported asphodel to their native daffodil, allegedly called the former "bastard 'affodil." "Jonquil" comes from the Spanish *jonquillo* (rush), referring to the rush-like leaves. Daffodils may have been brought to Britain by the Romans, who believed their mucilaginous sap could heal wounds, although in fact it contains sharp crystals that prevent animals from eating the plant and may in fact irritate the skin. But John

Parkinson says, "Know I not any in these days, with us, that apply any of them as a remedy for any griefe, whatsoever Gerard or others have written." Parkinson was right, for the sap of daffodils contains crystals of calcium oxalate, an irritant, which is why, in a vase of mixed flowers, daffodils will soon make the other blossoms wilt. The asphodel also has sharp crystals in its sap which protect it from being eaten, although its roots, unlike the daffodil's, are said to be edible and were used in times of famine by the Greeks.

❧ Narcissus, who was exquisitely beautiful, saw his own image in a pool, leaned over to possess it, and drowned, becoming the flower.

All three were associated with the dead. In Greek myth, pale asphodels grew in the meadows of the Underworld, kingdom of the dead. Hades abducted Persephone after she had wandered away from her companions to pick the flowers. The stupefying quality of their sweet perfume was once thought to be as dangerous as any narcotic, and many people find the scent overpowering. The Victorians suspected narcissi of having harmful "effluvia."

The name "narcissus" is most often associated with the Greek youth Narcissus, with whom the nymph Echo fell in love. He spurned her, and she hid in a cavern where she died of a broken heart, leaving only her voice. Meanwhile Narcissus, who was exquisitely beautiful, saw his own image in a pool, leaned over to possess it, and drowned, becoming the flower. People do love themselves when they think they love another, but they don't change into flowers—which was often a handy solution to a problem in Greek mythology.

The daffodil, for many, is spring itself. Describing the daffodils she and her brother William saw on a walk, Dorothy Wordsworth said, "Some rested their heads on these stones as on a pillow." This is good to remember when looking at daffodils after a storm: they are simply resting their heads. Dorothy noted that the daffodils "tossed and reeled and danced, and seemed as if they verily laughed with the wind, they looked so gay and glancing." One can't help wondering if William read her diary before writing his famous poem and wandering lonely as a cloud.

The daffodil has filled the time of as many poets as botanists, and almost everywhere people have traveled they have taken daffodils. Oscar Wilde said, "They are like Greek things of the best period," which is a way of saying that nothing really surpasses them and if possible they should be taken wherever we go—even to the Underworld.

❧ The Romans believed the daffodil's sap could heal wounds.

DAHLIA

BOTANICAL NAME: *Dahlia*. FAMILY: *Compositae*.

Dahlias are called after Dr. Anders Dahl, a Swedish botanist. Until recently they were also called "georginas," after the botanist Johann Georgi of Petersburg. The name is still used in Eastern Europe.

The history of their introduction is confused as well. They originated in Mexico and were grown by the Aztecs, who called them *"cocoxochitl."* Spanish invaders sent them home to the Old World, but dahlias did not, like some floral imports, take European gardeners by storm.

One story recounts that dahlia tubers were stolen from the royal gardens in Madrid and taken to the Jardin du Roi in Paris. Another nice story says that they were imported directly to France by a Monsieur Menoville, who had been sent to Mexico by the French government to smuggle out cochineal insects (a precious source of red dye, protected by the Spanish). Menoville reputedly sent the tubers home to Paris as food for the insects on the journey. The cochineal insects died, but the tubers were then sent on to the Jardin du Roi, whose curator, André Thouin, saw the dahlia as a possible edible substitute for the potato. Although it is not the proper food for the cochineal insect, the dahlia is said to be edible and the Aztecs had indeed used it for food. One Victorian described dahlias as having a "repulsive, nauseous peppery taste [which] inspires equal disgust to man and beast."

After Thouin's brief interest in the dahlia as a food source, the plant seems to have disappeared until several decades later. There had been no place for them either in the French formal gardens or the great English landscaped estates of the eighteenth century. But in the early nineteenth century seeds were sent to Berlin, where they were named for the botanist Johann Georgi; they were also sent by Lady Bute, wife of the British ambassador in Spain, to England, where they were named after Dahl, a physician and a pupil of Linnaeus. When they returned to the New World, they were known as "Mexican georginas."

The Empress Josephine uprooted her precious dahlia cultivars after some were stolen from Malmaison by a lady-in-waiting; we are not told what happened to the lady-in-waiting. Soon dahlias became immensely popular. In 1826 a prize of one thousand pounds was offered for a blue dahlia, and one dahlia tuber was reputedly exchanged for a diamond. Victorians enjoyed a showy lack of discretion in their material surroundings, and the contemporary style of gardening now fitted dahlias admirably. They now could connect the fashionable new shrubberies with the formal beds of flowers raised in hothouses and "bedded out." Perhaps this flamboyance was a way of compensating for discretion in so many other spheres of Victorian life.

DAISY

COMMON NAMES: Daisy, marguerite, ox-eye daisy.
BOTANICAL NAMES: *Bellis perennis, Chrysanthemum.*
FAMILY: *Compositae.*

Chaucer, writing about the English daisy, said that there was no "English rhyme or prose / Suffisant this floure to praise aright." Its botanical name comes from the Latin *bellus*, "beautiful." The English name comes from the Old English *daeges-eaye* or "day's eye," referring to the way the flowers open and close with the sun.

Flowers that open or close at certain times of the day were perhaps more noticed by our ancestors than by us, with our surfeit of timepieces. Andrew Marvell observed that the garden "computes its time as well as we," and Linnaeus actually made a floral clock whose flowers could accurately show the time throughout the day, according to when they opened. English daisies would not have been good in this respect, as they do not open on cloudy days.

Ox-eye "daisies," or marguerites (which are really chrysanthemums), may have got their name from their pearly color, from the Greek *margaretes* (pearl), or, some say, from the name Margaret.

Margaret of Anjou is the most likely candidate for this honor, and she had daisies embroidered on her personal banner. She was the wife of Henry VI who, in 1422, succeeded to the thrones of both England and France. Henry was supposed to have been saintly and ineffectual, but Margaret was ruthless and extremely ambitious to obtain the throne for her son. In the end both Henry and their son died and Margaret was exiled to France.

We associate daisies with simplicity, but the composite blooms are tube-shaped floral groups surrounded by petals, not one simple flower. An insect attracted to a composite pollinates dozens of flowers at once. The ox-eye daisies, which came with the colonists to America, were loved by poets and hated by farmers because their roots give off toxic substances that damage crops even as they fill great expanses with their beauty. Luther Burbank crossed them with a Japanese "daisy" to obtain, in 1890, the famous Shasta daisy. Apart from the Burbank potato it was his most successful hybrid and substantiated his belief that "man can modify, change, improve, add, or take away from any plant he chooses." He called the new daisy after the magnificent Mount Shasta near his home. It has remained a garden favorite ever since.

The daisy continued its tradition of modesty, and in Victorian times it was a popular name for sweet young girls. But although beautiful, like Margaret of Anjou, marguerites, which are the "daisies" grown in American gardens (the climate is not right for English daisies), have another side, just as she did. In a vase they will make the other flowers wilt. But a field of them looks like a sky studded with millions of stars—stars whose beauty for once is not inaccessible and that we can reach for and hold in our hands.

DATURA

COMMON NAMES: Angel's trumpet, thorn apple.
FAMILY: *Solanaceae.*

Daturas, most often seen in elegant New York vestibules, are expensive, showy plants with a sinister history. The botanical name comes from the plant's Arabic name *"tator,"* or its Indian name, *"dhat."* Indian thugs used it to poison their victims, and it was officially used to execute criminals. Linnaeus did not want to use the "barbaric" (Indian) name for the plant, so he modified it to the Latin root of *dare* (to give), because datura was given to those whose sexual powers were weakened.

Datura was one of the powerful ingredients of witches' ointments, rubbed onto their thighs and genitals to induce trances in which they soared above the world. Native Americans employed it more gently as an anesthetic and narcotic medicine. In South America, slaves and wives were given it before being buried alive with their lord and master. The herbalist John Parkinson called daturas "Thorne-

Apples" and warned that "the East Indian lascivious women performe strange acts with the seed . . . giving it [to] their husbands to drinke" (he does not elaborate on the acts). Thomas Jefferson said he avoided datura and other poisonous plants because "I have so many grandchildren and . . . I think the risk overbalances the curiosity of trying it." He goes on to say that during the Reign of Terror after the French Revolution "every man of firmness carried it constantly in his pocket to anticipate the guillotine. It brings on the sleep of death as quietly as fatigue does the ordinary sleep."

The datura of elegant atriums is closely related to jimsonweed or "Jamestown weed," so called because soldiers sent to Jamestown to quell Bacon's Rebellion in 1676 ate datura leaves, thinking they were salad greens. They were intoxicated for eleven days, reportedly sitting stark naked like monkeys, blowing feathers in the air, tossing excrement, kissing and gibbering. Nathaniel Bacon, who believed in unlimited territorial expansion, had led an unauthorized expedition against the Native Americans and taken over much of Virginia. He and the rebellion died that year, so the soldiers' intoxication did not turn out to be historically important, but the name of Jamestown stuck with the plant.

> Indian thugs used it to poison their victims, and it was officially used to execute criminals.

Datura contains scopolamine, used these days as a remedy for motion sickness—useful in perilous seas or even when really flying. Otherwise it welcomes those who enter tall buildings, on business of their own.

DAYLILY

COMMON NAMES: Daylily, tawny lily, lemon lily.
BOTANICAL NAME: *Hemerocallis*. FAMILY: *Liliaceae*.

The botanical name comes from the Greek *hemera* (day) and *kallos* (beauty) because the flowers' beauty lasts but a day, which is also why they are called "daylilies." They were named by Linnaeus, and the names *"fulva"* for the tawny lily and *"flava"* for the lemon lily are rare instances where he named specific plants by the color of their flowers.

Daylilies were used as food and medicine in China and Japan. They were dried or pickled in salt or cooked as vegetables. The flower buds of the *esculenta* variety were called *gum tsoy* (golden vegetable). The plants came to Europe early, possibly like rhubarb (also a medicinal plant), brought by traders along the silk routes from China. The Romans used them medicinally. The young leaves when eaten are said to be slightly intoxicating, and the Chinese had called the daylily *hsuan t'sao*, or the "plant of forgetfulness," as it was supposed to help allay sorrow by causing forgetfulness. By the time John Parkinson mentions them as "Asphodels with Lilly flowers . . . or Lillies with Asphodel rootes" which came from Germany, he says they "have no physicall use that I know, or have heard."

Daylilies were a popular garden plant in North America in colonial times and soon escaped to grow along roadsides. They are now bred enthusiastically and there are hundreds of varieties available. The mammoth tetraploid daylilies, much prized by collectors, are created with the help of colchicine, a substance used for manipulating plant genes. Colchicine is an extract first isolated in the 1820s from the autumn crocus, or colchicum, by two French scientists. It prevents cell division by inhibiting elongation of microtubules (the threads that pull chromosomes apart into opposite corners of a cell), so the cell does not divide and will contain twice as many chromosomes as a normal cell. Mammoth plants and flowers with these new large cells develop. Recently, however, there has been a trend toward more modest and fragile blooms, as well as new colors.

Although they are not native American flowers, daylilies are so much a part of the wild landscape, so easy to grow, and so rewarding that they are often used in fashionable modern "wild" gardens. Purists might disapprove, but daylilies are as much a part of America now as that other immigrant, apple pie. Both, thank goodness, are here to stay.

❧ Purists might disapprove, but daylilies are as much a part of America now as that other immigrant, apple pie.

DEUTZIA

BOTANICAL NAME: *Deutzia*. FAMILY: *Hydrangeaceae*.

Although deutzia was first described in 1712, it was not imported to Europe until the end of the nineteenth century. Its stems are hollow but they do not seem to have been used for flutes or pipes as other hollow-stemmed plants were. In fact, the deutzia doesn't seem to have any poetical associations at all. It's reliable, handsome, and a pleasure to have around, like many respectable lawyers with whom we are acquainted and who make good neighbors.

It is, in fact, called after a lawyer, Johann van der Deutz of Amsterdam. He seems to have been reliable, and maybe he was handsome as well. He was a town superintendent, an alderman, and a councilor. Together with David ten Hove and Jan van de Poll he provided money for Carl Peter Thunberg to investigate the natural history of South Africa, Java, and Japan (see "Japonica"). In gratitude Thunberg dedicated his *Flora Japonica* to them and named the genera *Deutzia*, *Hovenia*, and *Pollia* after them.

Deutz corresponded with the long-lived botanist and explorer

Joseph Banks, who, like Deutz, was born in 1743 (see "Everlasting Flower"). The banksias, which bear his name, however, are far less commonly grown than deutzias, and Johann van der Deutz, who was carried off in his forties, has in a sense outlived Banks in gardens, for many gardeners know him by growing his namesake.

Deutzias are native to China but long cultivated in Japan, where their wood was used for bodkins and cabinets and their leaves as furniture polish. Englebert Kaempfer, who first saw and described the Japanese deutzia, was employed by the Dutch East India Company as a doctor at their Deshima Island base. He taught the island's Japanese interpreters astronomy and mathematics in exchange for botanical specimens (although he knew they risked their lives by giving them), and he started a botanical garden on Deshima (see "Japonica"). He accompanied the Dutch embassy to Tokyo to pay respects to the emperor and, although closely guarded, managed to collect plant specimens along the way.

Kaempfer taught the island's Japanese interpreters astronomy and mathematics, in exchange for botanical specimens.

Deutzias have curving branched stems, covered with double white blossoms in June. They need very cold winters or they will flower prematurely, so they do better in the northern United States than in Britain. They are extremely beautiful and really should have poetry written about them, like other no more lovely plants. But that happens, doesn't it? Sometimes the most unworthy subjects can inspire extraordinary art, while the lawyer next door, full of grace, gets only a respectful obituary.

DOGWOOD

BOTANICAL NAME: *Cornus*. FAMILY: *Cornaceae*.

Dogwood was supposedly used to build the Trojan Horse. *Cornus mas*, or cornelian cherry, was valued by the Greeks for its exceptionally hard wood, used to make javelins and spearheads. John Parkinson said, "The wood . . . is very hard, like unto horne, and thereof it obtained the name" (from *cornus*, Latin for "horn").

How it became "dogwood" has to do with its edible and medicinal qualities. The berries of the *Cornus mas* are said to be edible and were supposedly fed to Odysseus's men when they were changed into pigs by Circe. *Cornus sanguinea*, or English dogwood, was called by John Parkinson "the Doggeberry tree, because the berries are not fit to be eaten, or to be given to a dogge." The Victorian garden writer John Loudon said that it was named because a decoction of its leaves was used to wash fleas from dogs, and L. H. Bailey said in 1922 that it was used to bathe "mangy dogs."

The American eastern dogwood, *Cornus florida*, is believed by some to be yet another "Crucifixion" tree (the tree on which Christ was crucified), although it is not native to the Middle East or Europe. Its bracts, leaf-like organs that look like flower petals, are shaped like a cross and at the base of each is a brown stain, like a blood spot made by a nail. Dogwoods need cold winters to set flowers, but late frosts will ruin the spectacular bracts that surround the cluster of insignificant yellow flowers. The American Pacific dogwood, or *Cornus nuttallii*, has four to six bracts, and they are not dented at the top like those of the eastern dogwood. It is named after Thomas Nuttall (see "Larkspur"). Late-flowering Asian dogwoods such as the Kousa dogwood of Japan were introduced to America and Europe at the beginning of this century. They do not flower until their leaves are out, whereas the American dogwoods suddenly ornament bare branches with a mass of papery blooms.

❧ The berries of the *Cornus mas* are edible and were supposedly fed to Odysseus's men when they were changed into pigs by Circe.

At different times, dogwood leaves, berries, and bark have been used to intoxicate fish, make gunpowder, soap, and dye (used to color the Turkish fez), make ink, and clean teeth (the twigs if chewed first will separate into a primitive toothbrush). Bark of the dogwood tree contains small amounts of quinine and "it is possible to ward off fevers by merely chewing the twigs" (Bailey). According to Peter Kalm, American settlers believed so strongly in the power of the dogwood that when cattle fell down for want of strength the settlers would "tie a branch of this tree on their neck, thinking it [would] help them." He does not comment on whether this helped or not, but he does say that "it is a pleasure to travel through the woods, so much are they beautified by the blossom of this tree." That, at least, is still true.

EVENING PRIMROSE

BOTANICAL NAME: *Oenothera*. FAMILY: *Onagraceae*.

All evening primroses originate from the American continent. They came to Europe in the seventeenth century and were called "primroses" because their flower resembled the yellow spring primrose (or "first rose") of Britain. John Parkinson called it the "Tree Primrose of Virginia" and said, "Unto what tribe or kindred I might referre this plant, I have stood long in suspense." Now we know that there are about 124 species of evening primrose, which form a "tribe" of their own.

As early as 1729, the Quaker gardener John Bartram had twelve different kinds of evening primrose growing in his botanical garden near Philadelphia. His garden formed the first collection of American native plants, and he traveled all over the East Coast to find them. He sent hundreds to the Quaker gardener and botanist Peter Collinson in London, and their correspondence is a delight to read.

Bartram was king's botanist and responsible for most of the American plant introductions of his time to England, but his contemporary Linnaeus called no flower after him. Later a sandpiper, *Bartramia*, was named for him, and a moss, *Bartramia*, was named for his son William, who loved birds and was not a gardener. John said, "I took no perticular notice of mosses but looked upon them as A cow looks at a pair of new barn doors yet now I have made A good progress in that branch of botany which realy is A very curious part of vegitation." The honors of nomenclature, as we have seen, are not necessarily appropriate to their namesakes.

Evening primroses do not all open in the evening. Most are recognizable by a cross-like stigma across the top of the style, which John Goodyer called "the nailes of the inner parts." The same formation is shared by the passion flower and was used by missionaries as an allegory to illustrate the Crucifixion. The evening primrose could have been used for the same purpose but seems to have come quietly to Europe with no religious, allegorical, or even medical associations. Since oil of evening primrose is sold nowadays by almost all health food stores, with quite extensive curative claims, it is surprising that in an era of plant medicine, Parkinson dismissed it, saying he "never knew any amongst us to use [it] in Physicke."

Its botanical name is from the Greek *oinos*, "wine," and *thera*, "to hunt." Etymologists seem to think that this name came from another, now unknown, Greek plant that was used to stimulate the appetite for wine. Why there should be a need to stimulate the desire for wine they do not explain. In any case, both the roots and leaves of the evening primrose are said to be edible and somewhat resemble parsnips in taste. Maybe they need a good wine to accompany them.

EVERLASTING FLOWER

COMMON NAMES: Everlasting, strawflower.
BOTANICAL NAME: *Helichrysum*.
FAMILY: *Compositae*.

A preoccupation of medieval scholars was discovering the "Riddles of the Queen of Sheba" not specified in the Bible. One popular riddle seen in medieval tapestries was of the queen showing King Solomon two flowers and asking him to guess which was the real one and which the artificial one. The king in his wisdom gets hold of some bees, which fly immediately to the real flower. The real flower, it is assumed, is soft-petaled and fragrant. Everlasting flower would have complicated the riddle beautifully if the queen or the scholars had thought of it. It doesn't look "real" at all, but has straw-like petals and no perfume except, after it dries, a smell that is supposed to repel moths.

The Egyptians knew the helichrysum, as did Pliny and Dioscorides, who say that the flowers, which last indefinitely, were used to decorate statues of gods. The Greeks made wreaths of the flowers and used them mixed with honey to treat burns. They named the plant for *helios* (the sun) and *chryson* (golden).

The Oriental helichrysum came to Britain from Crete in 1619 via the Padua botanic garden. John Parkinson grew it in his garden, calling it, charmingly, "Golden Flower gentle." He said it was "called by our English Gentlewomen, Live long and Life everlasting, because of the durabilitie of the flowers in their beautie."

The *Helichrysum bracteatum* from Australia is the one most often grown in our gardens now, and it is particularly well suited to dry conditions. It was brought back by Sir Joseph Banks, who was on Captain Cook's expedition to observe the transit of Venus and to seek out the *Terra Australis Incognita*, or "Unknown Southern Land" thought to "balance" the land mass of the Northern Hemisphere. Banks was one of the few who survived the voyage. When they were shipwrecked in Australia, Banks noted that the "almost certainty of being eat" on shore added to the unpleasantness. But he returned home and was director of the Royal Botanic Gardens for forty-two years. He was the instigator and patron of a great era of botanical exploration and sent Francis Masson, Archibald Menzies, Joseph Hooker, Clarke Abel, and David Nelson on their trips.

It is used nowadays in winter bouquets of dried flowers which rival silk flowers in popularity.

The everlasting flower was popular with the Victorians to make fireplace screens and otherwise decorate their hot, stuffy parlors where houseplants often couldn't survive. It is used nowadays in winter bouquets of dried flowers, which rival silk flowers in popularity. By the middle of winter both get rather dusty, and it is doubtful if even Solomon could get a bee to choose between them.

FORGET-ME-NOT

BOTANICAL NAME: *Myosotis*. FAMILY: *Boraginaceae*.

Its botanical name comes from the Greek *mus* (mouse) and *otis* (ear). This is from the rather touching perception that the leaves are shaped like a mouse's ears. John Gerard called it "scorpion grass" and believed that it cured scorpion bites, though there are no scorpions in England—but maybe it was best to be prepared.

The name "forget-me-not" comes from the Old French *ne m'oubliez mye*, which in turn was a translation of the German *vergiss mein nicht*. The best known legend about the flower is of a German knight picking a posy of forget-me-nots for his beloved as they strolled together on a riverbank. He slipped and fell in, but before drowning he threw her the flowers, crying, "*Vergiss mein nicht*." This excruciating story could really only have merit were it to be sung onstage with a suitably distraught and bosomy soprano and some excellent trap-door mechanisms. Botanically it doesn't hold much water.

Blue is a celestial color and it almost always clothes the Virgin

Mary in medieval paintings. Flies reputedly will avoid blue rooms, which is why dairies were often painted blue. Gardeners treasure blue flowers, which are rarer than any other color in our borders. Blue and yellow are also the colors that most attract insect pollinators, which is what interested Christian Sprengel in forget-me-nots. Sprengel, rector of Spandau in Germany, was investigating the role that colors of flowers play in the process of insect pollination. He so neglected his pastoral duties to pursue botany that he was dismissed from his post. In 1793 he published *The Newly Revealed Mystery of Nature in the Structure and Fertilization of Flowers*, which demonstrated his belief that all nature had a connected purpose. It was, however, unenthusiastically received by his contemporaries, leaving him too depressed to publish anything more. It was not until 1841 that Charles Darwin read the book and recognized the truths in it, which he incorporated into his own research.

> ❧ "Forget-me-not" is one of the few flower names that almost everyone knows and remembers.

"Forget-me-not" is one of the few flower names that almost everyone knows and remembers, and the flowers commonly decorate Valentine cards and the like. They grow in damp places and are, indeed, bluer than the Virgin's robe. The most memorable place they grew, however, was in Lady Chatterley's pubic hair, where her gamekeeper lover planted them, saying, "There's forget-me-nots in the right place." When she looked down at the "odd little flowers among the brown maiden-hair," she said, "Doesn't it look pretty!" The gamekeeper's reply is unforgettable: "Pretty as life," he said.

FORSYTHIA

COMMON NAMES: Forsythia, golden-bell.
BOTANICAL NAME: *Forsythia*. FAMILY: *Oleaceae*.

The Scottish gardener William Forsyth was a showy character, like the shrub that bears his name. After Robert Fortune (see "Bleeding Heart") had brought forsythia back from China and it had become popular, its ease of propagation and hardiness caused it to be planted in gardens everywhere. Then, like a lot of wildly popular plants, it fell into disrepute, and so-called discriminating gardeners talked of it as "vulgar."

Forsyth too fell into disrepute. In 1770 he became the director of the Chelsea Physic Garden. He was able and enthusiastic, reorganizing and replanting the Chelsea garden, exchanging seeds and plants with gardens abroad, making the first British rock garden with forty tons of old stone from the Tower of London and lava brought back from Iceland by Sir Joseph Banks, and helping to found the Royal Horticultural Society. However, in spite of all this laudable horticultural activity, he seems to have been a bit of a rascally entrepreneur. He invented, or claimed to invent, "Forsyth's Plaister." By 1799 overuse of forests had left few great trees suitable for wartime shipbuilding, and those that remained were often diseased. In his gardens, Forsyth had used his plaster to seal wounds in fruit trees after he had removed diseased limbs, and he offered to sell the recipe to the British navy. The navy fell for it and the treasury paid him fifteen hundred pounds—an immense sum in those days. The secret recipe turned out to consist of cow dung, lime, wood ashes, and sand mixed to a malleable paste with soapsuds and urine. Its efficacy was challenged by Thomas Knight, an expert on the cultivation of fruit trees, who refused to concede that "man, with the aid of a little lime, cow-dung and wood ashes, is capable of rendering that immortal, which the great God of nature evidently intended to die." A Quaker doctor, John Lettsom, supported Forsyth, but when challenged by Knight with a wager of a hundred guineas that he could not "produce a single foot of timber restored after being once injured to the state asserted by Mr. Forsyth," replied

❧ The secret recipe for "Forsyth's Plaister" turned out to consist of cow dung, lime, wood ashes, and sand mixed to a malleable paste with soapsuds and urine.

primly that his religion did not allow him to make wagers. Forsyth, however, died in 1804 before the controversy could be resolved.

Of course Forsyth may have believed that his plaster was truly effective—or it may even have worked. His instructions included cutting away the diseased parts of the tree before applying the cure and, as we are discovering increasingly, plants as well as people have self-healing powers we do not fully understand. The trees, with the canker removed, may simply have recovered as they would have without the plaster.

The forsythia asserts itself every spring with brilliant blasts of yellow, even sometimes where the house it adorns has fallen into ruins. Occasionally this bold showiness is devastated by an extra early frost, but mostly forsythia gets away with it, and cheers us all up with its very audacity.

FOXGLOVE

COMMON NAMES: Foxglove, fairy-bells, ladies'-thimble. BOTANICAL NAME: *Digitalis*. FAMILY: *Scrophulariaceae*.

Foxgloves, native to Britain and Europe, have always been considered fairy flowers. There are dozens of fairy names for them, as well as some more sinister ones like the Gaelic *ciochan nan cailleachan marblia*, or "dead old women's paps." In 1542, Leonhard Fuchs named the foxglove *Digitalis* from the Latin *digitus* (finger).

The name "foxglove" comes from the Old English *foxes glofa*, and the flowers do look like the fingers of a glove. Foxgloves tend to grow on woody slopes where foxes' burrows are often found. Foxes are wily creatures who may have needed magical gloves when they slunk out of the shadows and spirited away chickens. English foxes were brought to America in the eighteenth century by hunting club purists in Philadelphia, but soon those foxes interbred with native foxes. Foxglove plants were imported to America in the eighteenth century too, after their medicinal properties had been discovered, and they naturalized somewhat, but they do not fill the woods as they do in Britain. Foxes were protected for hunting in Britain and, like the flower, are commoner over there than here.

Foxglove was known to local healers like George Eliot's character Silas Marner, who gave an old woman foxglove tea "since the doctor did her no good," with miraculous results. This scene may have been inspired by William Withering, a Birmingham physician and member of the Lunar Society (a scientific club that met on the nearest Monday to each full moon), who in 1785 wrote *An Account of the Foxglove* in which he described how he had cured a patient of dropsy with foxglove tea using a recipe he had obtained from a healing woman. Erasmus Darwin (grandfather of Charles) belonged to the same club and later tried to claim credit for the discovery of digitalis, a potent cardiac stimulant made from the plant's seeds and dried leaves. It is still used today. Of the old women who first used the formula we know nothing. In 1787, Caspar Wistar (see "Wisteria") wrote that "dropsies are so often fatal we must try everything" but cautioned that one of Withering's patients in Edinburgh had died after being given foxglove tea.

Foxgloves are beautiful as well as useful. William Curtis, whose illustration of a foxglove was the frontispiece to Withering's book, compared the flowers to spotted wings of butterflies, which "smile at every attempt of the Painter to do them justice." The common ones are purple, but foxgloves now come in many colors. They are biennial but will self-seed profusely if they like where they are planted.

FUCHSIA

COMMON NAME: Lady's ear drops. BRAZILIAN NAMES: *Thilco, molle cantu* (beautiful bush). FAMILY: *Onagraceae.*

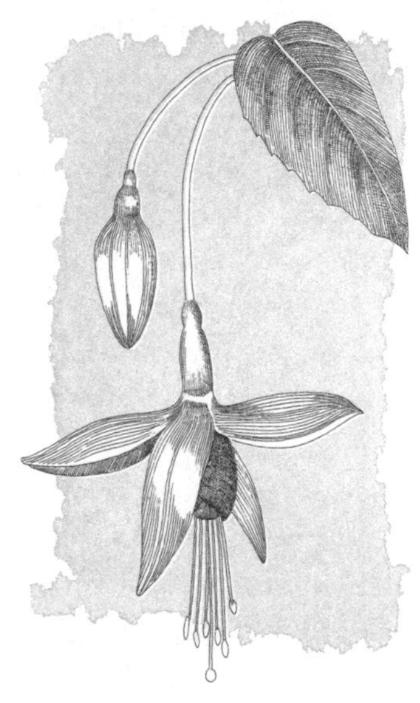

If Gertrude Jekyll, the expert on garden color, had designed the fuchsia, she might not have come up with so flashy or wonderful a flower, because fuchsias, like Chinese mandarins, do not know that scarlet and purple should not be juxtaposed. They are an exception to other rules too. They look more like bright plastic jewelry than flowers and are well named "lady's ear drops." They have "about eight" stamens, unlike most flowers whose constant number of stamens Linnaeus depended upon for his system of classification.

Most fuchsias originate in South America. The botanical name refers to the opinionated Leonhard Fuchs, a doctor at the University of Tübingen, whose summary of medicinal plants, *De Historia Stirpium*, written to correct "illiterate" botanists, turned out to be widely successful. In 1703 Charles Plumier (see "Begonia"), published *Nova plantarum americanarum* and in it first described the *Fuchsia coccinea* named in tribute to Fuchs (see also "Lobelia").

It is not clear exactly how fuchsias came to Britain in the 1700s, but they were first sold by James Lee, a Quaker, owner of the Vineyard Nursery, which specialized in exotic plants. In 1831 the *Lincoln Herald* published an account of Lee's supposed acquisition of the fuchsia. He was told of a plant in Wapping with "flowers hung in rows like tassels, their colour the richest crimson, in the centre a fold of deep purple." Lee "posted off" at once to ask the owner if he could buy it. "Ah sir," said the lady, "I could not sell it for no money, for it was brought me by my husband, who has now left again and I must keep it for his sake." The story goes on to say that Lee offered eight guineas for the "loan" of the plant—which he took back to his nursery where he pulled off "every vestige of blossom and blossom bud." He then divided and redivided it into cuttings until by the next year he had three hundred of them. He gave the original owner two plants, and sold the rest for a guinea each!

Fuchsias are popular houseplants in America and the more exotic varieties are not hardy. Some hardy fuchsias were so successful in British gardens that in the 1890s Canon H. N. Ellacombe described "houses covered with them from the ground to the roof, with spaces cut out for the windows."

As James Lee discovered, fuchsias are easy to raise from cuttings, and nurserymen have made plenty of money from them. Even so, they can never be commonplace, retaining always the exotic individuality which made Lee rush, posthaste, to that windowsill in Wapping.

GARDENIA

BOTANICAL NAME: *Gardenia*. FAMILY: *Rubiaceae*.

John Ellis, who was a merchant and botanist, insisted that Linnaeus call the gardenia after his friend, Dr. Alexander Garden. Ellis first suggested that Linnaeus call the Carolina allspice after Garden, but when the gardenia, or "Cape jasmine" as it was then called, arrived in London, he wanted that named for him instead. Linnaeus objected, saying that an American plant, discovered by Garden, would be more suitable, and that he was being criticized already for naming plants after his friends. Ellis simply told him that Dr. Garden already thought the plant had been named in his honor. And so it was.

Linnaeus was right about its unsuitability, because the gardenia, far from being American, came from South Africa. It is a member of the same family, Rubiaceae, as the madder, whose root yields a red dye, but its closest relative is the randia, a tropical shrub named for

Isaac Rand. The gardenia was discovered, in 1754, by Captain Hutcheson of the *Godolphin*, en route home to England from India. Hutcheson said he had gone for a walk on shore, become aware of a sweet, heavy scent, turned around, and seen a mass of huge, double white flowers. Even if you know what you are getting, a gardenia in full bloom is breathtaking. It is hard to imagine what it would be like if it seemed to have dropped from Heaven. Hutcheson dug it up and took it back to London. This time carrying the vision home worked; the plant survived and was propagated.

Dr. Garden was English but settled in Charleston, South Carolina, in 1751. He worked as a physician, collected plants, and kept in close touch with European botanists. He was opinionated as well as brilliant. He seems to have been a skilled and compassionate doctor, selflessly vaccinating patients during a smallpox epidemic, but he could be scornful, describing his fellow citizens as being as stupid as oxen and asses, and he objected to John Bartram's appointment as king's botanist because, he said, Bartram could hardly spell. Garden, a Tory, eventually had to leave America during the Revolution, and his granddaughter (who was named "Gardenia") was never allowed to meet him.

❧ Garden, a Tory, eventually had to leave America during the Revolution, and his granddaughter (who was named "Gardenia") was never allowed to meet him.

Before he left America, Garden was sent two gardenia plants by

Ellis; however, one died on the voyage from England, and the other died soon after. Gardenias are known to be tricky to grow. They are fussy about temperature and attract bugs. In 1767 Peter Collinson wrote to John Bartram about his gardenia, or Warner's Jessamin, "Who could think that fine plant had travelled so soon to your world . . . this Engaging Vegitable exercises the skill of all our Naturalists and yett I dont know any one has hit its Culture, for a year or Two it seems prosperous and then flags and Declines." Dr. Garden said he took the death of his plants as a bad omen for the perpetuation of his name. If, like the plant, his name did die out he said he would "make myself happy in some other acquisition if it should only be, like the former, imaginary." However, although they are tricky, people continue to grow gardenias—the reward of their beauty being worth the struggle.

GERANIUM

BOTANICAL NAME: *Pelargonium.*
FAMILY: *Geraniaceae.*

The name "geranium" was in use long before the flowers we usually call "geraniums" were known to the West. The garden geranium or cranesbill was named by Dioscorides from the Greek *geranos* (a crane), referring to its long, beak-like seed pod, similar to a crane's beak. Dioscorides listed it as a medicinal plant, and its many forms, still called "geranium," grow in the wild and in our gardens. What the rest of us sloppily call "geraniums," pedantic horticulturalists and real gardeners refer to as "pelargoniums," a genus of plants found originally in South Africa.

The South African plant was first called "geranium" by Jan Commelin, director of the Hortus Botanicus in Amsterdam and one of the three Commelins for whom the day-flower was named by Linnaeus (because two were prominent, like two of the flower's three petals, and one was inconspicuous and died "before achieving anything in Botany"). Soon geraniums started coming from the Cape in large quantities, and naming began to be a problem. In 1772 Francis Masson sent hundreds of pelargoniums back to Kew, calling them "geraniums." The director of Kew, Sir Joseph Banks, said, "We are principally indebted to Mr. Masson" for our geraniums, and indeed half of all known kinds were introduced by him. Other than this testimony, though, Masson got very little credit and suffered a great deal. He was chased through the African brush by a chain gang of escaped convicts, captured by the local militia to fight the French in Grenada, nearly killed in a hurricane off St. Lucia, and captured by French pirates on his way to North America. Eventually he made his way to Canada, where he is thought to have died on Christmas Day of 1805. He has no known grave. He did all this for a salary of one hundred pounds a year, and the only plant named after him is the massonia, a rare lily of which very few have ever heard.

By 1787, when Charles Louis l'Héritier de Brutelle (see "Gloxinia") published *Geraniologia*, there were so many South African geraniums that he invented a new genus for them, which he called *Pelargonium*. This means "stork's beak," from the Greek *pelargos* (a stork). He divided the geranium family into three: the cranesbills, which retained the name *Geranium*; the erodiums (from the Greek *erodios*, "a heron"), which are rock plants; and the pelargoniums, which are the South African geraniums.

Nursery catalogs, like A. A. Milne's dormouse, continue to call pelargoniums "geraniums"—except that the dormouse called them "geraniums (red)" and the catalogs call them "geraniums (pelargoniums)." Their proper name isn't exactly buried in an unknown grave, but it is seldom used, which is much the same thing.

GLADIOLUS

COMMON NAMES: Gladiolus, glad; *also (historical)* corn-flag, corn iris. BOTANICAL NAME: *Gladiolus.* FAMILY: *Iridaceae.*

The beautiful and the edible still tend to be divided by gardeners. We enjoy potatoes but would never, like Marie Antoinette, wear a corsage of their flowers. We grow millions of gladioli for their flowers but never think to eat their corms, which are said to taste like chestnuts when roasted and were certainly eaten in Africa, where many of them originated.

Before the African gladioli were common in the West, the Mediterranean and the rare British gladioli had been grown in gardens and used medicinally. John Gerard called them "Corne-flagge" or "Sword-Flag." They were known in ancient Greece and some scholars think they were the original hyacinth because the wild Greek gladioli have markings on their petals similar to those on the hyacinth (see "Hyacinth"). The name "gladiolus" comes from the Latin *gladius* (a sword), from the shape of the leaves. An ancient name for the gladiolus (and the iris) was "*xiphium*," from the Greek *xiphos* (a sword). John Parkinson, with his usual vivid accuracy, described the "stiffe greene leaves, one as it were out of the side of another, being ioyned together at the bottome." By the time he was writing, new gladioli had already been imported, and Parkinson said, "John Tradescante assured me, that hee saw many acres of ground in Barbary spread over with them." It must have been a splendid sight. Barbary was the Mediterranean region of Africa, and gladioli are not what we hear about there nowadays.

By far the largest number of our modern gladioli come from South Africa. From the end of the eighteenth century they were imported in huge quantities, including many sent back by James Bowie, a disreputable adventurer who botanized with Francis Masson (see "Geranium"). Too many introductions and hybridizations of gladioli have been made to enumerate here, but one important one was made in 1820 by Robert Sweet, whose career as a hybridist ended when he was accused of stealing garden pots from Kew. Another was the 'Maid of the Mist,' sent home in 1904 by Francis Fox, the engineer who built a cantilever railway bridge over the Zambesi River at Victoria Falls. This gladiolus was found flourishing in the waterfall's misty spray and had adapted to the constant moisture by developing a hooded upper petal which kept its pollen-bearing stamens dry. It introduced yellow and orange shades into the hybridized gladioli's color spectrum.

In most of Britain and North America gladioli have to be dug up and stored over the winter. Of course even if the climate is warm enough to leave the corms in the ground, mice share none of our compunctions about eating what is beautiful and can destroy a bed of gladioli as effectively as the severest freeze.

GLOXINIA

BOTANICAL NAME: *Sinningia speciosa.*
FAMILY: *Gesneriaceae.*

Gloxinias are not really gloxinias. True gloxinias are little-known plants from Brazil and Mexico of which there are about six species. We shall concentrate instead on how *our* gloxinias got their name.

It started in 1784 when Charles Louis l'Héritier de Brutelle, a French botanist, received a "true" gloxinia plant and named it as a compliment to Benjamin Peter Gloxin, who was a physician and botanical writer from Colmar. L'Héritier was the patron of Pierre Joseph Redouté, who illustrated much of his work, and was a respected botanist and wealthy man. He was entrusted with the herbaria of Joseph Dombey, which, in 1786, he took to London for "safe-keeping"—a move that provoked harsh criticism from some of his contemporaries. Dombey had been under the protection of the Spanish government on a botanical expedition to Peru. When he finally returned to France, the Spanish government demanded a large share of his collection, which the French would not relinquish, so l'Héritier took it to England and Redouté joined him there. Both l'Héritier and Dombey became victims of the French Revolution. Dombey fled and died in jail in the West Indies; when l'Héritier returned from England, he was imprisoned, then released to live in poverty, then assassinated.

In 1817, our gloxinia, which appeared to be similar to the true gloxinia, was introduced and named *Gloxinia speciosa*. Nurseryman Conrad Loddiges first described it, calling the new gloxinia a "most splendid subject." In 1825 things became more complicated. A new genus was established when a botanist at the University of Bonn, Christian Nees von Esenbeck, received a Brazilian plant and named it *Sinningia* after a gardener at the University, William Sinning. Then, in 1848 the genus *Ligeria* was established (called after a botanical author, Louis Liger), and it was decided that *Gloxinia speciosa* (our gloxinia) belonged in it, to distinguish it from the "true" gloxinia. Its name was changed to *Ligeria speciosa*. But in 1873 botanists decided that the genus *Ligeria* belonged in the genus *Sinningia*, and our gloxinia now became a *Sinningia* (*S. speciosa*), which it stayed. Except that everybody continued to call it a gloxinia!

❧ Gloxinias are not really gloxinias.

If it were not a popular plant, botanists might have used its new botanical name and no one would have known much about all these changes. But since gloxinias are hothouse favorites botanists have had to compromise. Gloxinias are sometimes called *Gloxinia (Sinningia) speciosa* even in horticultural dictionaries.

HOLLYHOCK

BOTANICAL NAME: *Alcea*. FAMILY: *Malvaceae*.

The common name comes from "holy" plus *hoc*, "mallow." It may have been "holy" because it was brought back to Britain by the Crusaders, and it was possibly called "hock leaf" because it was used to reduce swelling in horses' hocks, but it has been grown and used for so long that it is hard to be sure of the origins of its name. Herbs found in the fifty-thousand-year-old grave of a Neanderthal man included the remains of hollyhocks.

The botanical name of the hollyhock is from the Greek *alkaia*, or "mallow." Its relative, the marsh mallow, belongs to the genus *Althaea*, from the Greek *althaia*, "a cure." Garden hollyhocks and marsh mallows are of the mallow, or malva (Latin for "mallow") family. The wild marsh mallow was used medicinally; the root contains a mucilaginous juice said to be very soothing and that could be chewed by teething babies. Parkinson said hollyhocks "helpe to make the body soluble."

The mallow is of the same edible family as the hibiscus, cotton, okra, and rose of Sharon. Marsh mallow root was used to make the confection called marshmallow, but, while people nowadays might worry about the edibility of the original plant, they do not seem to worry about the edibility of its sticky namesake.

Albertus Magnus, the medieval botanist and theologian who traveled over Europe compiling an encyclopedia, recommended that hollyhocks should be rubbed on the hands to protect from burns during "ordeal by fire," which was a popular method of trying criminals at the time. Nicholas Culpeper, who wrote *The English Physician* in 1652, said that hollyhocks were good for "Belly, Stone, Reins, Kidneys, Bladder, Coughs, Shortness of Breath, Wheesing, Excoriation of the Guts, Ruptures, Cramp, Convulsions, the King's Evil, Kernels, Chin-cough [whooping cough], Wounds, Bruises, Falls, Blows, Muscles, Morphew [a skin eruption], Sun-burning." In addition to all this their fibrous stems can be used to make cloth, and they yield a very good blue dye.

No wonder that hollyhocks were one of the earliest imports to America. Colonists brought both the *Alcea rosea*, the old single red hollyhock, and the *Althaea officinalis*, the marsh mallow, and gave their seeds to the Cherokee Indians soon after they arrived.

❧ The wild marsh mallow was used medicinally; the root contains a mucilaginous juice said to be very soothing and that could be chewed by teething babies.

In the eighteenth century new strains with bigger, double flowers were brought from China; hollyhock became one of the most popular garden flowers, the standby of cottage gardens until the nineteenth century, when hollyhock rust came to Britain. At that time the recommended cure was to remove infected leaves, but now it can be controlled with modern sprays. It's rather sad though, if the whole point of hollyhocks was their curative powers. They are pretty flowers in the border, but not quite the same if they have to be sprayed with poisons.

HONEYSUCKLE

BOTANICAL NAME: *Lonicera*. FAMILY: *Caprifoliaceae*.

Adam Lonicer, or Lonitzer, was a German physician who practiced in Frankfurt and, in 1555, published a book on natural history. A later botanist said, "His herbals though used widely and reprinted many times, had little or no influence on Linnaean and post-Linnaean taxonomy," but, in 1753, Linnaeus named the honeysuckle *"Lonicera"* after him. The bush honeysuckle, *Diervilla*, is of the same family (see "Weigela").

Wild British honeysuckle was known from earliest times as "woodbind," although other climbing plants bore the same name. Possibly Shakespeare's "woodbine" is convolvulus (see "Morning Glory"). Generations of children have "sucked honey" from the beautiful flowers, which John Gerard compared to the "nose of an Elephant." They are pollinated by the hawk moth, and a Viennese botanist, Kerner, once placed a hawk moth three hundred yards away from the nearest honeysuckle early in the day and marked it. When dusk fell, he watched the moth gently wave its feelers and then head straight for the honeysuckle.

Sometimes honeysuckle is trained around young straight branches, which are later cut to make walking sticks with a spiral imprint on them. But it always had to be controlled in gardens. John Parkinson said he knew the wild honeysuckle "yet doe I not bring it into my garden, but let it rest in his owne place, to serve the senses that travell by it, or have no garden." The "Northern" honeysuckle, he said, was "entertained into their gardens onely [by those] that are curious."

We should have been less curious when we entertained the Japanese, or Hall's, honeysuckle. It was introduced to America in 1862 by Dr. George Hall, an American doctor who opened the Hospital for Seamen in Shanghai. It was carefully sent home in a Wardian case and cosseted in a Long Island nursery, one of the treasures that the Parson's Nursery officials unpacked as carefully as "an original of Raphael or Murillo," but it soon spread to cover much of the eastern seaboard and has become a terrible pest.

When cursing Hall's honeysuckle we have to remember that he also introduced the beautiful little star magnolia. Meddling with plant distribution is, as we know now, tricky. As in love affairs, it is hard to get the right balance. We introduce plants, rejoice, and coddle them, but when they prosper we call them invasive. We should have left them alone—but then we would have missed much that is beautiful as well. Japanese honeysuckle is a menace, but on the banks of busy highways its scent even overwhelms the stench of exhaust.

HOSTA

COMMON NAMES: Hosta, funkia, plantain lily.
BOTANICAL NAME: *Hosta*. FAMILY: *Liliaceae*.

Hostas became popular in this century when gardeners were looking for easier ways to care for their gardens. They grow splendidly in almost complete shade and will form a mass of diverse and extraordinary leaves, some of them as big as large wet buttock prints. Their colors range from almost true blue to yellow, and they are wrinkled and striped, puckered, pointed, and blunt according to the dozens of kinds now available. Once they are planted, you don't have to do anything to them, except admire them.

The garden hostas originated in China and Japan. Engelbert Kaempfer (see "Deutzia") described them in 1712 and Latinized their Japanese name as *Joksan vulgo giboosi*. Carl Thunberg (see "Japonica") later called them *Aletris japonica* and *Hemerocallis japonica*. Hostas were named "funkias" by Christian Sprengel (see "Forget-me-not") as a tribute to Heinrich Christian Funck, a German botanist and specialist on mosses. The hosta, especially in Britain, is still sometimes called "funkia," or "funckia." Hostas are also called "plantain lilies," from the Latin *planta*, "sole of the foot," which the large leaves resemble.

> Hostas grow splendidly in almost complete shade and will form a mass of diverse and extraordinary leaves, some of them as big as large wet buttock prints.

No one seemed able to agree about what to call the plant. (Botanists do agree that the name by which a plant is first described, unless it is found to be incorrect, should be retained.) Many have yearned to have their names first stamped on a plant because once there, it is there forever, even if it is their only claim to eternity. The hosta fell short of their ideal. By 1812 its names included *Giboosi*, *Hemerocallis*, and plantain lily. That year, Sir Edward Salisbury divided hostas botanically and proposed the names *Niobe* and *Bryocles* for them, but he never published his description or validated the names. The same year Leopold Trattinick proposed the name *Hosta*, after the physician Nicolaus Thomas Host, an expert on grasses who seemed an acceptable recipient for the name of a lily. At first the new name appeared to be invalid, as a verbena (now known as *Cornutia*) already bore the name *Hosta*. But it turned out that the cornutia had been classified a verbena by Linnaeus, so finally, in 1905, the

International Botanical Congress voted that the name *Hosta* could be used.

The German eye surgeon Philipp von Siebold sent the first living hosta plants from Japan to Leyden Botanical Garden. The Japanese allowed only Dutchmen from the East India Trading Company on Deshima Island (see "Japonica"), and although Siebold spoke Dutch, he had a thick German accent that didn't deceive the Japanese. So he pretended his accent was that of a remote mountain town in Holland and was allowed to stay!

Even the Dutch were not allowed to leave Deshima for the mainland, but Siebold was a skilled surgeon and was permitted to leave Deshima to perform the first cataract operations in Japan. As payment for his services, he accepted botanical specimens, which he illicitly exported. He fell in love with a Japanese woman who registered as a prostitute in order to live with him on Deshima. They had a daughter. In 1829, he was caught shipping out a map of Japan and was arrested and expelled. Siebold pined for Japan and wore Japanese dress even in Ghent, where he died.

HYACINTH

BOTANICAL NAME: *Hyacinthus*. FAMILY: *Liliaceae*.

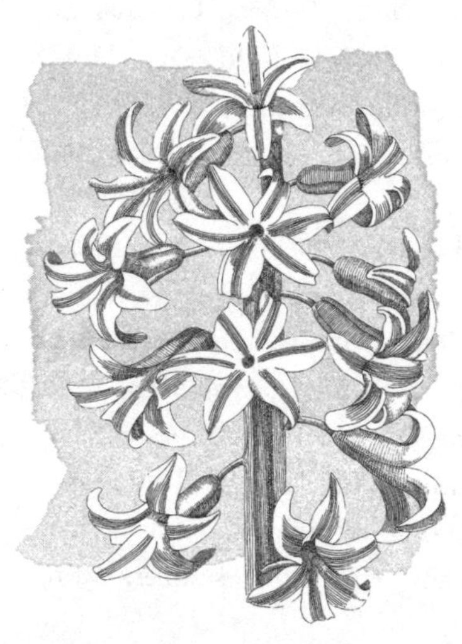

Hyacinth was a beautiful boy whom the god Apollo loved. While they were playing the ancient game of quoits together, Hyacinth ran forward to catch the discus, but it struck him on the head and killed him. (A chilling elaboration of this story tells that Zephyr, the wind, was jealous of the friendship and blew the quoit against a rock to rebound and kill the boy.) As Hyacinth died, a flower sprang from his bleeding head, which hung over the shoulder of Apollo, who was desperately cradling him in his arms and begging him to live. Wild hyacinths always bend toward the ground, and the letter-like markings on their petals were supposed to read *AI, AI* in Greek, the sound of a mournful wail. Wild gladioli and some wild orchids in Greece have the same symbol, and they were all used in wreaths for the dead. According to John Parkinson, hyacinths "hinder young persons from growing ripe too soone," which would be useful to those who love unchanging youth.

Apollo's ardor was also responsible for Daphne turning into a laurel bush and Cyparissus changing into a cypress tree. So he left a tree, a bush, and a flower—almost a complete small garden of metamorphosed passion. With a tub or two of impatiens, it would be enough for a townhouse garden—or indeed any garden if one considers the implications.

The hyacinth was brought early to Europe from Turkey and was grown in Europe's first botanical garden, in Padua. It was observed and probably collected by the German physician Leonhardt Rauwolf when he went to Turkey in 1573. Rauwolf wrote an account of his travels that was translated into English by the botanist John Ray. He is the first Westerner to describe coffee, which, he noted, made him feel "curiously animated."

The hyacinth quickly became a popular garden plant. Originally there were only four colors, but by 1725 there were two thousand named cultivars, including double hyacinths, which are rare in gardens today. Hyacinths seem to create passionate opinions, and some dislike their formal bearing and heavy scent. They can be forced to bloom in early winter, which some find cheering; others find a contrived spring of this kind unnerving and prefer to wait for the real event.

Catherine Morland, in Jane Austen's *Northanger Abbey*, informs Mr. Henry Tilney that she has "just learnt to love a hyacinth," although she is "naturally indifferent about flowers." He says, "But now you love a hyacinth. So much the better. You have gained a new source of enjoyment, and it is well to have as many holds upon happiness as possible." Of course they are not really talking about hyacinths at all. They are in the middle of a story about love, misunderstanding, and jealousy, but this time, it has a happy ending.

HYDRANGEA

COMMON NAMES: Garden hydrangea, hortensia.
BOTANICAL NAME: *Hydrangea macrophylla*.
FAMILY: *Saxifragaceae* until recently, now *Hydrangeaceae*.

The garden hydrangea has a wonderful quality of changing from pink to blue, according to the soil in which it is grown. We have wanted the power to change the colors of flowers ever since Pliny suggested soaking seeds or bulbs in wine to achieve this (see "Lily"). Garden hydrangeas will turn from pink to blue if the soil is acid and if aluminum is available to them, but it still seems rather magical, and when they were first introduced it was inexplicable. It was initially thought that they might take their color from their surroundings, especially as cuttings from a plant of one color might well turn out the other color when propagated. By the beginning of the nineteenth century it was suspected that they could be made blue by adding iron or alum to the soil, but in 1832 John Loudon said this was "not yet ascertained," and a mixture of sandy loam and fresh sheep's dung "produces the same effect."

The garden hydrangea puzzled botanists in another way. It was introduced in about 1788, and its large, showy flower head, consisting mostly of sterile flowers with only a few fertile ones, made it hard to classify what were petals, what were sepals, and what were flowers, according to the Linnaean sexual system.

The American hydrangeas came to Britain in the eighteenth century. The name "hydrangea," given by Linnaeus, came from the Greek *hydro* (water) and *aggeion* (vessel), and it is generally thought to refer to the cup-shaped fruits rather than to the fact that hydrangeas need a lot of water (a large one will take ten to twelve gallons a day in hot weather.

The garden hydrangea was named Hortensia by Philibert Commerson, who accompanied Bougainville on his voyage around the world in 1766 (see "Bougainvillea"). It is usually supposed that the name "hortensia" was after Mlle. Hortense, daughter of the prince of Nassau; the latter had joined Bougainville's expedition in order to escape his creditors. But it is worth noting that the woman named Jeanne Baret, who had sailed on the voyage disguised as a boy (called Jean), changed her name to Hortense when she settled in France. We will never really know why. Anyway, in 1830 the name was changed to *Hydrangea macrophylla* ("large leaved"), by which it is now known.

The big pink or blue garden hydrangea is as common in America as blue-haired old ladies, and has the same feel of dyed unreality. It always had slightly refined associations from the moment it arrived in England at the London docks and a special delegation, including the great Sir Joseph Banks himself, went down to meet it personally and hosted a breakfast reception in its honor! It clearly demands, and receives, more attention than other plants.

IMPATIENS

COMMON NAMES: Impatiens, busy Lizzie, touch-me-not. BOTANICAL NAME: *Impatiens*. FAMILY: *Balsaminaceae*.

Impatiens, from the Latin *impatiens* (impatient) is named from the way the seeds are jettisoned out of their pods. Erasmus Darwin (see "Foxglove"), in an appalling poem explaining the Linnaean system, said the impatiens "with rage and hate the astonished groves alarms / And hurls her infants from her frantic arms." John Parkinson said impatiens seeds "will soone skippe out of the heads, if they be but a little hardly pressed betweene the fingers." The plants propogate profusely because of their method of shooting out their seeds, and we're probably lucky that our garden impatiens are not frost hardy; otherwise we might scorn them as much as dandelions. Impatiens are in fact perennial, but the first frost turns them into limp black rags.

Jewel-weed, which is also an impatiens, grows wild, filling ditches and woods. Nobody values it much, even though the juice from the leaves soothes poison ivy rash, the boiled stems are edible, and the whole plant yields a good yellow dye. Another impatiens, policeman's helmet or *Impatiens glandulifera*, was carefully imported to England from the Calcutta Botanic Gardens in 1839, and has taken over English streams.

The impatiens that we drape ubiquitously, like joyful flags, over gardens and in planters at the first sign of spring is *Impatiens sultanii*, called "busy Lizzie" in England. John Kirk sent it to Europe from Zanzibar in 1865, and Joseph Hooker named it in honor "of that distinguished potentate, the Sultan of Zanzibar to whose enlightened philanthropic rule eastern Africa owes so much." It was later given the name *Impatiens walleriana* in honor of the Rev. Horace Waller, who was a missionary in central Africa.

John Kirk was British consul in Zanzibar. He had previously gone to Africa with the famous Dr. Livingstone. The hardships of that expedition were so terrible that Kirk came to the conclusion that "Dr. Livingstone is out of his mind." He lost eight volumes of botanical notes and one hundred drawings when their canoe overturned in rapids. Once they climbed over rocks so hot they were badly burned. Midges plagued them so thickly that the local people were able to press the insects into cakes, which, said Kirk valiantly, "tasted not unlike caviar."

The impatiens family is vast and botanically almost incomprehensible. Joseph Hooker, the famous botanist and director of Kew, was trying to sort it out when he died. He called it "deceitful above all plants" and "worse than orchids."

IRIS

COMMON NAMES: Iris, flag, gladdon *(ancient)*.
FAMILY: *Iridaceae*.

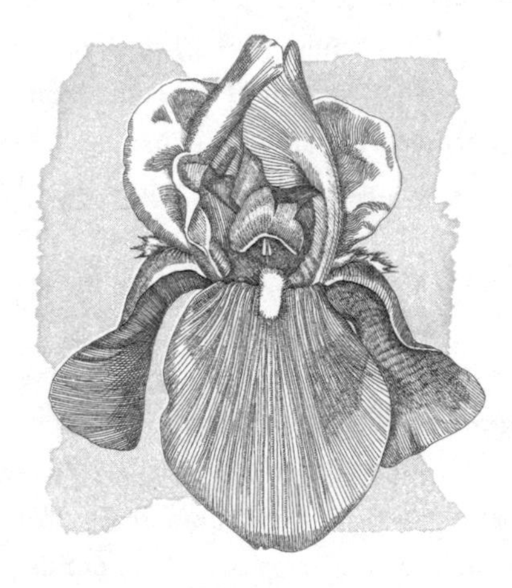

Iris was the messenger of the gods and the rainbow linking earth with other worlds. She escorted souls along her iridescent bridge to another life, and she herself used it to join the thoughts of gods and men. She was that longed-for connection to those whom we love intensely, but who are suffering without our awareness, and it was she who was sent to tell Alcyone, still praying for the safety of her husband Ceyx, that he had already drowned.

The "iris" is also what we call a part of the eye—into which we look in recognition. Eye color is one of the few human features that cannot be changed except by covering it. Maybe it too is a kind of bridge between the known and the unknown. There were irises carved on the temple at Karnak, and they were used to adorn gardens, graves, and the brows of Egyptian gods. White irises, the color of mourning, were planted on Muslim graves.

The flag iris is supposed to have saved the life of the sixth-century

Frankish king Clovis, who then succeeded in conquering much of France under the Christian banner. God, or common sense, showed Clovis, trapped by the Goths at a bend in the Rhine, flag irises growing where it would be shallow enough to cross the river and so escape. In gratitude he adopted the iris flower as his emblem, and it became the symbol for the kings of France. Irises were on Louis VII's banner during the Second French Crusade (1147) and were called *fleur de Louis*, which in turn became fleur-de-lis.

Irises were called fleur de Louis, which in turn became fleur-de-lis.

With its trinity of petals, the iris was an important religious flower, especially dedicated to the Virgin Mary. The leaves are sword-shaped, and some older dictionaries call the iris (and the gladiolus) "xiphium," from the Greek *xiphos* (a sword). The blade-like edges represented on one side the sharpness of Mary's pain at the sufferings of her son and, on the other, her sharp defense against the Devil.

Irises have been grown in Japan, China, Siberia, and almost all the temperate world. Tropical climates are the only places where they cannot be found. Although there are some native American irises, others were brought early to America from Europe. On May 26, 1811, Thomas Jefferson, writing to his granddaughter Anne about transition, said: "The flowers come forth like the belles of the day, have their short reign of beauty and splendor, and retire, like them, to the more interesting office of reproducing their like. . . . The Irises are giving place to the Belladonnas . . . as your mamma has done to you, my dear Anne . . . and as I shall soon and cheerfully do to you all in wishing you a long, long good-night."

JAPONICA OR FLOWERING QUINCE

COMMON NAMES: Japonica, flowering quince, cydonia. JAPANESE NAME: *Boke.* BOTANICAL NAMES: *Chaenomeles speciosa, Chaenomeles japonica.* FAMILY: *Rosaceae* (once *Pyrus,* or pear, then *quince*).

Japonica blossoms burst out of bare branches in earliest spring before there are green leaves anywhere. They are sometimes white, but more usually red or brilliant coral, and they seem more like an implausible statement against the darkness of winter than real flowers. Henry Reed, the World War II–era poet, uses their "silent eloquent gestures" in his poem "The Naming of Parts" as a statement against the darkness of war:

> Japonica
> Glistens like coral in all of the
> neighbouring gardens,
> And today we have naming of parts

The naming of the japonica itself is complicated. The first japonica was named by the Swedish botanist Carl Peter Thunberg, a pupil of Linnaeus. Thunberg learned enough Dutch to be allowed on Deshima Island as chief surgeon for the Dutch East India Company. Deshima was an artificial island on thirty-two acres in Nagasaki Harbor. No one who was not Dutch, or pretending to be Dutch, was allowed there, except about two hundred Japanese servants, interpreters, and prostitutes. The Dutch were not allowed to leave the island for the mainland, but Thunberg ingeniously bypassed this restriction by sifting through the hay brought for the livestock to Deshima and gathering enough new botanical material to enable him to write *Flora Japonica* in 1784. In it he described and named the thorny *Pyrus japonica*, or "Japanese pear."

After this, the japonica played for a while a kind of nomenclatural musical chairs. A thorny plant from China was first named *Pyrus japonica* and then renamed *Pyrus speciosa*, but was still popularly called "japonica." Meanwhile, the original japonica was renamed *Pyrus maulei*, or Maule's pear (Maule was a nurseryman in Bristol known for cultivating it). Neither plant was found to be a member of the pear family, and for a while they were classed as quinces and called "Japanese quinces" or "cydonias," named

❧ Japonicas are sometimes white, but more usually red or brilliant coral and they seem more like an implausible statement against the darkness of winter than real flowers.

for Kydon, in Crete, where the quinces were supposed to have originated. The poet Keats, in an 1819 letter to his sister, used the name "japonica" referring to a camellia that he described shading a globe of goldfish in a window looking out over the Lake of Geneva.

Finally japonicas came to rest botanically by being classed as *Chaenomeles*, from the Greek *chaina* (to gape) and *melon* (an apple), referring to a perception that the fruit was split. Thunberg's original plant and its descendants became *Chaenomeles japonica*, and the plant from China and its descendants became *Chaenomeles speciosa*. Both are more often called "japonicas" or "flowering quince." Both produce brilliant blossoms in early spring, followed by a hard pear or quince-like fruit that can be made into jelly.

JASMINE

COMMON NAMES: Jasmine, jessamine. BOTANICAL NAME: *Jasminum*. FAMILY: *Oleaceae*.

The name comes from the ancient Persian name for the plant, *yasmin*. The *Jasminum officinale* has been cultivated so long that it is uncertain how it came to Europe. The Persians valued it highly and knew how to extract its scent by steeping the blossoms in sesame oil. In England jasmine was used to cover arbors. Some people found the scent overwhelming, including Gilbert White, who wrote in his journal in 1783, "The jasmine, now covered with bloom, is very beautiful. The jasmine is so sweet that I am obliged to quit my chamber."

The Chinese winter jasmine, or *Jasminum nudiflorum*, is so called because the yellow flowers are borne on the naked winter branches. It was introduced in 1844 by Robert Fortune (see "Bleeding Heart") who compared the blossoms to "little primroses." The Chinese used them to make aromatic *Heung Pin* (Fragrant Leaves), which is green tea combined with dried jasmine blossoms.

The American or Carolina jasmine, *Gelsemium sempervirens* ("everlasting jasmine"), is of the Loganiaceae family, and isn't really a jasmine. Its name comes from *gelsomino*, Italian for "jasmine." It is evergreen with bold yellow flowers that are poisonous and have no fragrance. Thomas Jefferson grew it and planned to cover large tracts of unused garden with "Jessamine, honeysuckle, sweetbriar, and even hardy flowers which may not require attention." This was to be "an asylum" for wild animals except, he says "those of prey." He doesn't say how to keep out the animals of prey, or the poison ivy, but it is an alluring concept of cohabitation as lovely as democracy—and we are still working on both ideals.

The *polyanthum* jasmine, so called because it has many flowers (Greek *poly*, "many," and *anthos*, "flower") was sent from China by George Forrest, who had explored there for twenty-eight years. In 1905 he had survived a massacre when eighty missionaries and other Europeans were waylaid by nationalist Tibetans; all but fourteen were killed or captured. Forrest saved himself by rolling down the steep mountainside. He was hunted for eight days, discarding his boots to elude their dogs and living on a handful of dried peas that were in his pocket. He finally walked to safety over remote mountain paths, in spite of a sharp bamboo stake penetrating his foot.

The only drawback of jasmine is its floppy habit of growth, which John Gerard describes as a "need to be supported or propped up, and yet . . . [it] claspeth not or windeth his stalkes about such things as stand neere unto it, but onely leaneth and lieth upon those things." But he adds that jasmine is "good to be anointed after baths, in those bodies that have need to be suppled and warmed"—which surely includes the bodies of all gardeners.

KERRIA

COMMON NAMES: Kerria, bachelor's button.
JAPANESE NAME: *Yama buki* (Kaempfer).
BOTANICAL NAME: *Kerria japonica.* FAMILY: *Rosaceae.*

Many of our most cheerful flowers are called "bachelor's buttons." Is this because they manifest a prenuptial *joie de vivre*, or because they are considered symbols of availability?

Tudor and Stuart men wore clothes as elaborately embroidered with flowers as those of women. Even at the Battle of Edgehill, in 1642, Charles I wore a purple military sash embroidered with carnations, roses, and tulips. Men wore flowered clothes on and off until the nineteenth century, when embroidered waistcoats were still worn, one of the most famous being the waistcoat in "The Tailor of Gloucester," written and illustrated by Beatrix Potter, which was "embroidered with pansies and roses" and was only finished in time for the lord mayor's wedding by the nocturnal assistance of a mouse.

The introduction of real flowers worn on the lapel is sometimes attributed to Prince Albert, Queen Victoria's husband. The prince, in a passionately unfrugal gesture at the termination of his bachelorhood, was said to have taken a rose out of her wedding bouquet, split open his lapel, and stuck it in. Whether or not this is true, boutonnieres were commonly worn by Victorians, bachelors or not, the most famous being Oscar Wilde's chrysanthemum.

Kerria was introduced by William Kerr in 1805. Kerr was a young gardener sent out from Kew to China by Sir Joseph Banks, who especially instructed him to search for new fruit trees. Kerr was to do this on a meager salary of one hundred pounds a year. He sent as many plants as he could home on merchant ships, but there was a high rate of attrition in this era before the invention of the Wardian case (see "Bleeding Heart").

Kerr had other problems too. After his arrival in Canton, it was reported that he was "unable to prosecute his work, in consequence of some evil habits contracted, as unfortunate as they were new to him." He had probably become an opium addict. He is supposed to have introduced, among other things, the tiger lily and the nandina bamboo. In 1810, Banks recommended him as superintendent of the new botanical garden in Ceylon. He worked there from 1812 until his death in 1813. No one knows how he died; officially it was of "some illness incidental to the climate."

His legacy, the kerria, became an immensely popular plant, grown in cottage gardens all over Britain. Its flowers are clear yellow, double or single. They flourish with no care and, like many shrubs from China, do well in America. There's nothing like a carefree shrub, or a carefree bachelor for that matter, to cheer one up.

LADY'S MANTLE

BOTANICAL NAME: *Alchemilla*. FAMILY: *Rosaceae*.

The botanical name, *Alchemilla*, is from the Arabic *al kimiya*, meaning "alchemy," possibly referring to the land of Khemia (Egypt) where such arts originated. Extracting the juices of plants was said to be the first step in changing base metal into gold with the hope of obtaining untold riches, curing all diseases, and infinitely prolonging life.

Lady's mantle was an important magical plant. Unlike most plants, which get rid of their excess water as vapor, it exudes water during the night in the form of dewdrops around the rim of the leaf. These droplets, which a seventeenth-century herbal called "pearls that falleth in the night," were considered magical and added to the medicinal importance of the plant. It was thought to be a powerful aphrodisiac, giving "lust to the worke of generacyon" (Culpeper). John Gerard said it "keepeth down maidens paps or dugs . . . when they be too great and flaggie"—a problem difficult to cope with in the era before the bra.

It was called lady's mantle because of the shape of the leaves, which look like a soft green cloak. However, such a mantle was not necessarily seen just as a covering but also something that, if possible, a man might get under. Before books were available and widely read, plants and their shapes, habits, and flowers were often used as illustrations of life—and some were raunchy illustrations, full of sexual innuendos. Most of the "lady" flowers originally suggested the parts or possibilities of women. When Christianity arrived in Britain the early fathers converted flowers as well as people. "Lady" flowers were ingeniously transformed into "Our Lady" flowers—symbols of the blessed Virgin Mary. Lady's mantle was now the softly draped cloak wrapping the mother of God. What made this more appropriate was the fact that the flowers are parthenogenic and can set seed without fertilization—a fact not understood by early church fathers but maybe observed. When the Puritans came to power in England they disapproved of "Popish nonsense"—and "Our Lady's" flowers became "lady's" flowers once again. The name stuck, and when *Alchemilla mollis*, a bigger alchemilla and the one we grow in our gardens, was introduced from Turkey in 1874, it remained "lady's mantle."

❧ These droplets, which a seventeenth-century herbal called "pearls that falleth in the night," were considered magical and added to the medicinal importance of the plant.

LARKSPUR AND DELPHINIUM

BOTANICAL NAMES: *Consolida, Delphinium.*
FAMILY: *Ranunculaceae.*

The larkspur and its close relative, the delphinium, are both named from the shape of their flowers. The larkspur flower looks a bit like the claw of a bird, and the delphinium flowers, "especially before they be perfected" (Gerard), resemble the bottle-like nose of the dolphin, *delphis* being the Greek for "dolphin."

The first garden larkspurs, annual natives of Britain, were called *Delphinium consolida*, from the Latin *consolida*, "made whole," referring to the medicinal properties of the plant. Both it and the *Delphinium ajacis*, which was introduced to Britain about 1573, were thought to be efficacious against poisonous stings, and both plants, like all Ranunculaceae, are poisonous. They were used, dried and powdered, as very effective insecticides.

The *ajacis* part of the annual larkspur's name comes from the legendary Greek hero Ajax, who almost killed Hector by throwing a great stone at him. Achilles finally finished Hector off with a spear. Afterward, when Paris killed Achilles by shooting an arrow into his vulnerable heel, Achilles's armor was to be given to the most valiant of the Greeks remaining, and the choice was between Ajax and Odysseus. Minerva tilted the vote in Odysseus's favor, since she valued intelligence mixed with valor, and poor Ajax was not particularly bright. Ajax, completely dishonored, lost his reason temporarily and mistook the sheep around the camp for his rivals. He went berserk, killing sheep and beating to death a ram that he took to be Odysseus. When he finally came to his senses, he realized what he had done, and his only honorable way out was to kill himself. He flung himself onto his sword, and from the blood that fell to the ground sprang larkspurs. Their petals (like those of the hyacinth) are marked with the Greek letters *AI*, the Greek cry of mourning.

> As they were getting ready to defend themselves, the barrel of Nuttall's gun was found packed with dirt — he had been using it to dig up plants.

The American larkspur, *Delphinium nuttallianum*, was found by and named after Thomas Nuttall, who explored and botanized in Oregon and northern California. It was used by West Coast Indians to make a blue dye and by European settlers to make ink. Nuttall was known as "Le Fou" by contemporaries. He canoed down rivers even though he could not swim, and on one occasion his party was threatened with an Indian raid; as they were getting ready to defend themselves, the barrel of Nuttall's gun was found packed with dirt—he had been using it to dig up plants. He would probably have settled in America had it not been for his English uncle's bequest of an estate on condition that he live there, and he died in England. He had a beautiful garden at Nutgrove—where at least he would have been able to grow delphiniums. Even the annual larkspur does not last well in hot, humid American gardens, and perennial delphiniums often have to be renewed each year. We don't need them to make lethal powders either, for we've learned, too well, since the days of Ajax, to make better weapons against insects—and one another.

Their petals are marked with the Greek letters *AI*, the Greek cry of mourning.

LAVENDER

BOTANICAL NAME: *Lavandula*. FAMILY: *Labiatae*.

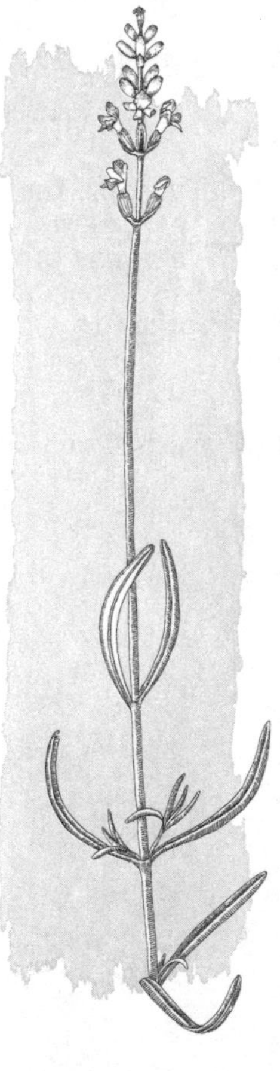

Washing (to the extent that all but certain small boys practice it nowadays) is fairly new in the West. But the name of lavender is not new; it comes from the Latin *lavare* (to wash). Lavender was used from ancient times to make perfumes and to scent such soaps as there were.

The Romans washed and took frequent baths in public bathhouses, but after the fall of the Roman Empire, the bathwater was tossed away with the rest of that civilization. Even royalty seldom washed; they used perfume liberally instead. Elizabeth I took a bath occasionally, but James I/VI never even washed his hands, which he "rubb'd" with the wet edge of a napkin. Water in sixteenth-century England was often contaminated with sewage and washing in it probably would not have been very healthy anyway. One of the lures of colonists to the New World was that the water was so pure it could even be drunk and "those that drinks it be as healthful, fresh and lustie, as they what drink beere" (Captain John Smith). Even so, somehow out of all this filth and pollution flowered some of the most beautiful literature that has ever been.

Soap, when available, was very expensive. In 1562, four pounds of gray soap cost twice as much as a whole pig (which was sixpence) and six times as much as a dozen eggs, but almost anyone could grow lavender, and it was so common that in 1568 the botanist William Turner said it "were but lost labor" to describe it. It was one of the cheaper perfumes, which were an important part of hygiene. Lady Macbeth, when agonizing over that bad "little hand" of hers, does not talk of soap and water but of "all the perfumes in Arabia." But of course she could have afforded something better than lavender water.

> Lady Macbeth, when agonizing over that bad "little hand" of hers, does not talk of soap and water but of "all the perfumes in Arabia."

Real perfumes were, as they are now, pretty expensive. Workmen handling frankincense in Alexandria were "sowed up and sealed" into their breeches so they could not conceal it in body crevices. Lavender water was easy to make, but pure oil of lavender was a luxury. It takes two thousand pounds of blossoms to make ten pounds of distilled lavender essence.

By the nineteenth century, soap and water had come into fashion and the use of perfumes was suspect. Henry Phillips, writing in the 1820s, called the use of perfumes in men an "effeminate practice" brought to Rome from

Greece and said "we would recommend the old practice of laying clean linen in lavender, in preference to throwing the extract of it on dirty clothes."

Although introduced to Britain early, lavender is probably native to the Mediterranean (some say it may originally have come from India). It likes chalky dry soil and bright sunshine, and although it can die back and grow up again from the roots in spring, it doesn't stand extreme cold.

It is good to grow lavender. The old writers said that it would "comforte the brayne very well" and that you can "imbibe good humour" from it. The herbalist John Gerard warned against its overuse by "unlearned Physitians and . . . foolish women," but said that it would "helpe the panting and passion of the heart." Whether this is true or not, a bed of lavender, or a handful of it in a drawer, is a comfort to the nose, the brain, and the heart. It's nearly as good as a hot shower.

LILAC

BOTANICAL NAME: *Syringa*. FAMILY: *Oleaceae*.

"Lilac" comes from the Arabic *laylak* or Persian *nylac*, meaning "blue." Funnily enough, the name "lilac" now means another color—more purple than blue. The botanical name, *Syringa*, is from the Greek *syrinx*, "a pipe," because the pithy stems can be hollowed out. The mock orange, or philadelphus, is sometimes called syringa too, and has smaller stems. Both were used by the Turks to make pipes, and they were introduced to the West at about the same time.

Ogier Ghiselin de Busbecq probably brought the first lilac back from Turkey, but it had already been described by Pierre Bélon, who had visited the court of Suleiman the Magnificent in Constantinople with an embassy party sent by Francis I. Bélon described strange wonders he saw abroad (even though at home in the French court the king liked to sleep with a lion on his bed) and he said the lilac flower was "like a fox's tail." The privet, of the same olive family, has similar flowers but their perfume is musky, and John Gerard accused this "white lilac" of "molesting the head in a very strange manner." Lilacs are sometimes grafted onto the stock of their sturdy privet cousins, and sometimes these hybrids die and a privet appears instead of a lilac—to be rooted up by the mystified gardener.

Lilacs and olives seem very distant relatives, but both are permanent anchors of civilization, rusty anchors sometimes in the seas of change. For both live forever and will remain long after the farmers who planted them have died. Olive trees, planted by people whose very language has disappeared, still cling to crumbling Mediterranean terraces. American settlers planted lilacs in front of farmhouse doors, not for usefulness but for beauty, while they struggled to make a new life in the wilderness. Sometimes the slowly cleared fields, the houses, and the walls were no more permanent than those who made them, but the lilacs remained by the ghost porches, leading nowhere.

Bélon's fox-tailed lilac tree was growing in Vienna in 1562, and soon the *Syringa vulgaris* spread wherever people traveled and lived, as it was hardy in all climates. At the end of the nineteenth century, during the Franco-Prussian War, Victor Lemoine introduced new double lilacs. Apparently Lemoine's sight was failing, so his wife had to pollinate the lilac flowers, standing on a stepladder. The Lemoine, or "French," lilacs are still widely grown today. During this century, new hardy lilacs came West from Asia. Among them is the delightful pink 'Miss Kim.' A 1987 survey showed that one in every five Koreans has the surname "Kim"—so this lilac honors many young ladies, the original one not now known, rather like a floral unknown soldier.

Even when lilacs die, they retain their perfume. For if you burn their wood, the sweet fragrance endures in the smoke, reminding you that, like all brave souls, lilacs are forever.

LILY

BOTANICAL NAME: *Lilium*. FAMILY: *Liliaceae*.

The lily's name may have pre-classical origins. But its Greek name, *leirion*, and its Roman name, *lilium*, gave us the name we know it by. The Romans were said to cure corns with the juice of the bulbs (although one wonders why they had corns if they usually wore sandals). In marriage ceremonies the lily was a symbol of purity which, along with wheat, the symbol of fertility, made an ideal matrimonial mix, still hard to acquire.

Whether or not the Madonna lily is the lily of the fields referred to in the Book of Matthew (see "Anemone") has fueled debates among biblical scholars, but medieval paintings of the Virgin do almost all include a lily. The lily was adopted by the Church as the Virgin Mary's flower because, unlike most flowers, its scent, similar to the "scent of sweetest frankincense," cannot be extracted as essential oil, and the flower reflects the purity and worth of Mary, being "white without and gold within." The stamens and pistils of the lilies on church altars were removed so they "remained virgin," and Madonna lilies were always white, in spite of Pliny's instructions for making them purple by soaking the bulbs in red wine.

> ❧ In marriage ceremonies the lily was a symbol of purity which, along with wheat, the symbol of fertility, made an ideal matrimonial mix, still hard to acquire.

We still have white "Easter" lilies, and florists remove the stamens, partly (they say) to prevent pollen making a mess and partly to make the gelded blooms last longer. Actually our "Easter" lily, although white, is not the true Madonna lily but one of the Oriental lilies (which come in many colors) introduced in the nineteenth century.

Ernest Wilson, called "Chinese Wilson" because he explored so extensively in China, just escaped sacrificing his life to lilies. He went twice to China, the second time in 1910, to collect the regal lily. He had gathered an enormous load of lily bulbs and was on his way home with them when his mule train was caught by an avalanche. He jumped out of his sedan chair just before it was hurled down a precipice. His leg was shattered by a falling rock. There was a mule train coming the other way, and the only way it could pass without, perhaps, causing another avalanche was for Wilson to lie on his back while more than forty mules stepped over him. He reached safety but was left with what he called a "lily limp." He died soon after when his car went over the edge of the road, not far from the Arnold Arboretum in Boston where he worked.

The regal lily is white within but wine-colored outside—perhaps a symbol of our lack of true purity where plants are concerned? Cowards, or the "lily-livered," were said to have a pale white liver—the liver, with the heart, being a source of human courage. Maybe wine red, like blood red, is the color of passionate—if misplaced—bravado, like Wilson's lying in agony on his back and counting mules' bellies so we could have lilies in our gardens.

LOBELIA

COMMON NAMES: Lobelia, cardinal flower.
BOTANICAL NAME: *Lobelia*. FAMILY: *Lobeliaceae*.

The blue lobelia we use most often in our gardens is *Lobelia erinus*, which was imported from South Africa in the 1800s and immediately became very popular as an edging or container plant. Its bright clear blue was much in demand for colored "ribbon" plantings. In the nineteenth century color was rediscovered like a new dimension. Bright colors and their juxtapositions interested not only gardeners but also artists and thinkers. Artists began painting out of doors using new pigment paints like rose madder, cobalt blue, and cadmium yellow, for the first time available in handy metal tubes. The theory of color became a preoccupation ranging from the bedding plans of gardeners to Goethe's studies.

Newly imported tropical plants from South America and Africa were used to make colored parterres, or geometric beds with different flowers outlining their design. Strong flower colors, like yellow, red, white, and blue, were in great demand. The *Lobelia erinus,* whose bright, clear-blue blooms last all summer and form low, thick carpets resistant to rain and wind damage, was included in most kinds of geometric gardening, and it still is. *Erinus* may be from the Greek *eri*, "early," meaning "spring flowering." But although the plant is perennial and will last through the winter if brought inside, we are more apt to plant it outside in spring and let the frost kill it.

Lobelia cardinalis is a North American plant —supposedly called *cardinalis* by Queen Henrietta Maria, wife of Charles I, because the color reminded her of a cardinal's robe. *Lobelia siphilitica*, also from North America, was described by Peter Kalm (see "Mountain Laurel") as a cure for the pox (syphilis). *Lobelia cardinalis* and *Lobelia siphilitica* are both perennials that grow well in American borders. They are a clear patriotic red and blue respectively.

❧ The *Lobelia erinus*, whose bright, clear-blue blooms last all summer and form low, thick carpets resistant to rain and wind damage, was included in most kinds of geometric gardening.

Lobelia was named by Charles Plumier (see "Begonia") for Matthias de l'Obel (see "Candytuft"), who corrected John Gerard's famous *Herball* (and accused him of pilfering material). He also attempted a new classification system of plants, subdividing them by leaf characteristics. He wrote a history of cereals, a description of roses, and instructions for brewing beer. His own name, Obel, comes from *abele*, the white poplar, and his family crest was *Cadore et Spe*, symbolized by the white undersurface of poplar leaves, demonstrating candor, and the green upper surface, hope.

• He wrote a history of cereals, a description of roses, and instructions for brewing beer.

Lobelia erinus comes in white and pink too. But like most good blues it really defeats its purpose to have it in other colors. Its wonderful blue fairly shimmers, and pots of it look like dished-up summer sky, or bowls of glinting Mediterranean sea that, even on dreary days on noisy town terraces, we can keep right there beside us.

LOOSESTRIFE

COMMON NAMES: Yellow loosestrife, purple loosestrife. BOTANICAL NAMES: *Lysimachia, Lythrum.* FAMILIES: *Primulaceae, Lythraceae.*

The original name of yellow loosestrife, or *Lysimachia vulgaris*, comes from King Lysimachus who was the companion and successor of Alexander the Great. Lysimachus's name came from the Greek *lysi machein*, "causing strife to cease," and loosestrife was used to prevent yoked animals from fighting (a sprig was put between them) and to staunch the wounds of war.

Purple loosestrife, called "lythrum," from the Greek *lythron*, "blood" (referring to the flower's color), grows in streams and boggy places. It was once thought to be the "long purples" near which Ophelia drowned, but scholars decided that those were really orchids. However, the famous pre-Raphaelite painting by John Millais of the drowned Ophelia shows her lying in a stream with loosestrife on its banks. Evidently Millais and his friend, Holman Hunt, looked a long time for an appropriate place for the tragedy and found it on the Ewell River in Surrey. Millais painted the streambed with its banks of loosestrife, leaving a gap for Ophelia, who was later inserted. The long-suffering model, Elizabeth Siddal, was painted lying in a cold bath in Gower Street. Siddal, like Ophelia, seems to have been unlucky in love. In 1860 she married Dante Gabriel Rossetti, another pre-Raphaelite artist for whom she modeled, but in 1862 she took an overdose of laudanum and died. Rossetti apparently felt he had neglected her and remorsefully buried his only complete book of poems with her. Seven years later, however, he had her body exhumed so he could publish the poems. As for Millais, he married John Ruskin's wife.

Purple loosestrife is important botanically because it is trimorphic. These kinds of plants bear flowers with three different lengths of stamens and pistils, so a bee can only cross-fertilize by matching together different configurations of each flower, and has to go to another plant to do so. Thus self-fertilization is impossible. This sophisticated system is one of the wonders of the botanical world and made Charles Darwin, who studied it, write to Asa Gray, "I am almost stark, staring mad over Lythrum."

Purple loosestrife blooms at the end of summer during the angry heat of August, but when the hope of coolness is near, they make sheets of brilliant color across steamy lowlands. It spreads in streambeds and marshes, both in Britain and the United States, and is regarded as a pestilential weed, although Mrs. William Starr Dana, in *How to Know the Wild Flowers*, says, "One who has seen the inland marsh in August aglow with this beautiful plant, is almost ready to forgive the Old Country some of the many pests she has shipped to our shores in view of this radiant acquisition."

LOVE-IN-A-MIST

BOTANICAL NAME: *Nigella damascena.*
FAMILY: *Ranunculaceae.*

The botanical name comes from the Latin *niger* (black). Its seeds are black, and the flower was used mostly for them, which in the Middle East were put in cakes and bread. According to the Bible scholar Harold Moldenke, "Egyptian ladies eat them to produce stoutness, which is considered an attribute to beauty in these lands." *Nigella damascena* is called after and is believed to have come from Damascus.

Its common name is associated with beauty too, but of a particular kind. "Love-in-a-mist" refers to the fine, hair-like leaves that surround the flowers. Evidently these fine hairs around the blossoms were suggestive. The French, less coyly, call them *cheveux de Vénus*.

The Western literary association of women with flowery images goes back to the Song of Solomon and has not always been ethereal, even if disguised in fragrant petals. In more primitive times close living quarters made refinement difficult, and dozens of flowers had openly sexual associations. Navelwort (or lady's navel) was used by the Romans with the hope that the juices of its leaves *cum vino circumutus* (mixed with wine) *in pudendis contrictionem laxat*. As living conditions improved, there was more room for delicacy in life, but flowers remained a useful way of referring to what gradually became less mentionable. By the eighteenth century ladies were openly and admiringly compared to flowers. Oddly, at the same time, Linnaeus was explaining the sexuality of the flowers themselves. He described the andromeda as "flesh-coloured" and added that "her beauty is preserved only so long as she remains a virgin (as often happens with women also)—i.e. until she is fertilized." No wonder his new system was dismissed by some as "lewd." Linnaeus himself said that the sexual attributes "of plants we regard with delight, of animals with abomination, of ourselves with strange thoughts" (see "Rudbeckia").

❧ Linnaeus himself said that the sexual attributes "of plants we regard with delight, of animals with abomination, of ourselves with strange thoughts."

For the Victorians, names such as Rose, Violet, and Daisy were immensely popular as emblems of innocence and freshness. The irony was not lost on writers. Young ladies wandering through heavily perfumed conservatories as innocently as flowers, like George Eliot's

Maggie Tulliver in *The Mill on the Floss*, sometimes got more than they bargained for. Maggie is in the conservatory during an interval at a ball, when Stephen, who is with her and watching her reach for "a large, half-opened rose," is seized with a "mad impulse" and showers kisses onto her wrist. He is confronted by a furious Maggie, who asks how he dares "insult" her. Maybe it was easier to walk the line between delicacy and sexuality when ladies were safely sequestered, even though the flowers in the garden might be named for the thighs of an aroused nymph (*'Cuisse de Nymphe Emue'*—see "Rose"), a maiden's hair, or even a priest's pilly. Ellen Willmott, who never married but who owned Warley Place and had about a hundred gardeners at her beck and call, knew what she was saying when she pointed to a pink rose and said, "That is Cupid: I knew him not."

LUPINE

COMMON NAMES: Lupine, lupin, bluebonnet.
BOTANICAL NAME: *Lupinus*. FAMILY: *Leguminosae*.

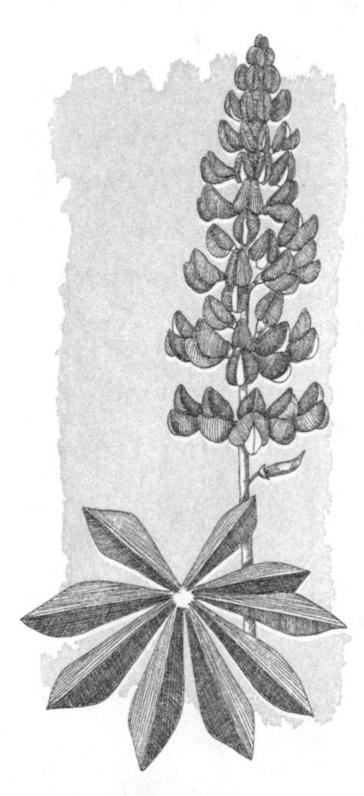

Like wolves, lupines, from the Latin *lupinus*, were supposed to ravage the land and destroy it. Actually they are good for the land as, like all legumes, they fix nitrogen in the soil. They were also called *Pisum lupinum*, because their peas were said to be fit only for wolves. These peas or seeds are very bitter until boiled several times, but quite nourishing. The Stoic Zeno compared himself to lupine seeds because when well soaked (with wine) he was less bitter. Those seeking to communicate with the dead at the Oracle of Epiros were fed a diet of lupine seeds, which induce a state of intoxication, perhaps making such communication more accessible.

The Mediterranean lupine is an annual. Perennial lupines are from North America and were introduced to Europe in the seventeenth century. In 1748 Peter Kalm, traveling round North America (see "Mountain Laurel"), noticed that livestock ate "almost all other plants save the lupin," although the leaves were green and "extremely soft to the touch." Charles Darwin studied lupine leaves and found they "sleep in three different manners" when they close at night. In the daytime they are in constant motion, sometimes rotating ninety degrees to follow the sun.

There is a legend that bluebonnets, the Texas wild lupines, were brought to the New World by the Spaniards, whose Crusaders had originally taken them from the Holy Land. There is no biblical record of lupines, but bluebonnets are a distinct limited species that seem celestial when in flower. Henry David Thoreau, describing a hillside of bluebonnets, said, "The earth is blued."

> Like wolves, lupines, from the Latin *lupinus*, were supposed to ravage the land and destroy it.

In spite of being in gardens so long, lupines were not really popular until they were hybridized by George Russell. His long life was spent as a jobbing gardener, and he did all his work in a small council allotment, relying entirely on insect pollination and simply pulling out less desirable flowering plants. In spite of his fame in the horticultural world he retained his pragmatic approach to life, refusing always to donate his flowers to funerals. He maintained they were for the living, not the dead. His own funeral was simple and his grave unmarked, but his memorial lives on in gardens everywhere each summer.

MAGNOLIA

BOTANICAL NAME: *Magnolia*. FAMILY: *Magnoliaceae*.

When we look into a magnolia flower we are looking back into prehistory. Magnolias were among the first plants on earth to reproduce using flowers pollinated by insects. Some prehistoric plants, such as water lilies, survived worldwide. Others, such as the ginkgo, survived in a very limited area. The magnolias were found in Asia and America but not in Europe.

The earliest magnolias to reach Europe came from America. Francisco Hernandez, physician to Philip II of Spain, described a Mexican magnolia in the 1570s but, as far as we know, the first live magnolia—a Virginia or swamp magnolia—was sent to Europe by John Banister (see "Bluebell") in 1688. The Virginia magnolia is interesting because of the swiftness with which it indicates it has been pollinated. Within half an hour the flowers wither and go brown.

The *Magnolia grandiflora* was growing in Britain by 1737, when the famous botanical artist Georg Ehret, a German who lived in England from 1736 until his death in 1770, walked three miles every day, from Chelsea to Fulham, to study its opening buds in Sir Charles Wager's garden. He drew it for Mark Catesby's *Natural History of Carolina*, although most of the illustrations were by Catesby himself (see "Silver Bell").

Far outshining the American magnolias, however, were the magnolias introduced from China, the first of which was sent to Sir Joseph Banks at Kew in about 1780. They were much hardier than the magnolias from America (which were grown outdoors in England but needed a sheltered position), although they were at first thought to be greenhouse plants. By this time the genus had been named, some say by Charles Plumier, some say by Linnaeus, after Pierre Magnol, Louis XIV's doctor and a professor of botany at the University of Montpellier.

One of the most popular magnolias found in gardens today, both in Britain and America, is the pink *Magnolia* × *soulangiana,* which appeared in 1826 in the garden of Monsieur Etienne Soulange-Bodin, who founded the National Horticultural Society of France. In the garden, which was near Paris, there were two magnolias, *M. denudata* (or *heptapeta*) and *M. liliiflora* (or *quinquepeta*), and the chance seedling was a cross between them. Monsieur Soulange

❧ Monsieur Soulange had been tutor to the Empress Josephine's children, a witness to her civil marriage to Napoleon, and superintendent of her garden at Malmaison.

had been tutor to the Empress Josephine's children, a witness to her civil marriage to Napoleon, and superintendent of her garden at Malmaison. He was also a patron of the artist Redouté. It is said that Napoleon disliked him.

It is not known if Pierre Magnol was acquainted with magnolias, but they were well named for him: he was the first to divide plants into "families," and they are certainly a distinctive family. William Bartram recognized magnolias in 1791 as an extraordinary and "celebrated family of flowering trees," and we are still celebrating their ancient and particular beauty.

• Magnolias were among the first plants on earth to reproduce using flowers pollinated by insects.

MARIGOLD

COMMON NAMES: Pot marigold, French marigold, African marigold. BOTANICAL NAMES: *Calendula officinalis, Tagetes.* FAMILY: *Compositae.*

Marigolds are associated with the sun. They bloom under blazing skies when everything else is wilting and you don't care if it does. The pot marigolds even open and close with the sun. "Such is the love of it known to be toward that royall Star," said Thomas Hyll in *The Gardener's Labyrinth*, that "they . . . at the noon time of the day fully spread abroad, as if they with spread armes longed . . . to embrace their Bridegroom."

The name comes from "Mary's gold," for marigolds were flowers of the Virgin Mary and were used to decorate church altars. The original European marigold was called "calendula," from the Latin *calendae*, "the first day of the month," because it bloomed every month of the year in ecclesiastical and monastery gardens, constantly supplying flowers for the Church.

The calendula is also called "pot marigold" because it was put in cook pots and was often used as a cheap substitute for saffron to color cakes, butter, and puddings. John Gerard says it was used for "broths, physical potions and for divers other purposes." These included hair dye, as the sixteenth-century herbalist William Turner sternly explained: "Some use to make their heyre yelow wyth the floure of this herbe, not beying contēt with the naturall colour which God hath geven them."

The popular marigolds we plant in American gardens are *Tagetes*, named by Linnaeus and otherwise called "French" or "African" marigolds, although both originated in South America, probably brought back from the New World by the Spaniards and then taken to France. Charles V brought marigolds back from a crusade, calling them *flos Africanus*.

The name comes from "Mary's gold," for marigolds were flowers of the Virgin Mary and were used to decorate church altars.

The *Tagetes* are called after Tages, the grandson of Jupiter. He taught the Etruscans haruspicy, which is the useful art of foretelling the future by examining entrails. He was a god of the underworld who rose out of a ploughed field with this bright idea, which was a standby in ancient forecasting. The Greeks were always stopping to examine entrails before deciding on a new move, much as we listen to weather forecasts or stockbroker predictions (perhaps with comparable results). Anyway, the introduction of so brilliant a skill earned for him the name of a brilliant flower—at least Linnaeus must have thought so when he chose that name for the marigolds.

The roots of *Tagetes* marigolds exude thiophenes, which kill nematodes, and they are often planted in vegetable gardens to repel pests. Early gardeners said they repelled vermin.

We associate marigolds with the heat of summer and dazzling sunshine. If they are the Virgin Mary's flower, they would certainly turn toward the light. But even gods of the Underworld must be enticed by it too—or they would stay in the entrails of the earth where they belong. Whomever's flower they are, marigolds add a blaze of brightness to our gardens and to our lives too, as they search for, and digest for us, the fiery brilliance of the sun.

❧ The calendula is also called "pot marigold" because it was put in cook pots and was often used as a cheap substitute for saffron.

MONTBRETIA

COMMON NAMES: Montbretia, crocosmia.
BOTANICAL NAMES: *Crocosmia* × *crocosmiiflora, Tritonia* × *crocosmiiflora*. FAMILY: *Iridaceae*.

Montbretias are to Britain what orange daylilies are to America: anyone who has any sort of garden can, and does, grow them, and even if there is nothing else in the garden they are sure to be there. Most fancy garden books leave them out altogether. After all, like common daylilies, they are orange, which is not considered a tasteful color—especially as Nature has provided so many purply pink flowers that are apt to juxtapose themselves right next to them and distress the poor gardeners who are trying so hard not to be vulgar.

Montbretias are called after Antoine François Ernest Conquebert de Montbret, one of the botanists who accompanied Napoleon on his expedition to conquer Egypt in 1798. It went well at first. Montbret and Alire Raffenau-Delile, the chief botanist, studied the sacred lotus. Redouté painted. They made a botanical garden in Cairo. They even created a menagerie in a villa garden. The captain of artillery, Boussard, found and took French possession of the famous Rosetta stone. But, as unfortunately often happens, the cultural triumph needed military security as well. The British fleet under Lord Nelson destroyed the French fleet at Abukir, and one disaster followed another. Napoleon escaped and went back to France, leaving the botanists stranded. Finally the botanists who were left after the fighting were allowed to go back to France with a shipment of plants that Raffenau-Delile would not abandon. The British admiral admired his pluck and let him take them home, but the Rosetta stone was taken by the British and is still in the British Museum.

"Crocosmia" comes from the Greek *krokos*, or "crocus," and crosme, "smell," because the dried flowers smell like saffron.

Montbretias are actually a hybrid of tritonias and crocosmias, both of which come from South Africa and which are very closely related. "Crocosmia" comes from the Greek *krokos*, or "crocus," and *crosme*, "smell," because the dried flowers smell like saffron. "Tritonia" comes from the Greek *triton*, "weathercock," because the stamens face in different directions. The famous French plant breeder Victor Lemoine (see "Lilac") first crossed them in 1880 to get the montbretia. This turned out to be hardier than either parent, and it flourished to excess in British gardens. It is also called *Crocosmia* × *crocosmiiflora*, which literally and idiotically means "crocosmia crossed with flowers like crocosmia." It reminds one of a European monarch's family tree, but far from being debilitated by convoluted genealogy it has taken over Britain and is increasingly found over here, sometimes sold as montbretia and sometimes as crocosmia. However montbretias won't survive in colder regions and have to be dug up and stored, so the daylilies are probably safe.

Crocosmias were found by William Burchell, who was an eccentric English plant explorer. In 1810 he gave up his post as a schoolteacher and traveled round South Africa at his own expense. He lived and traveled in an oxen-drawn covered wagon from which he flew the British flag. He took no notice of warnings about dangerous territories and would take no companion with him. He stopped occasionally and played his flute to any natives who would listen—and apparently was so bizarre that even in areas highly dangerous to foreigners he was left unharmed. He was passionate about flowers, noting in his diary that he had "feelings of regret that at every step my foot crushed some beautiful plant." His diary also included notes on the best way to cook ostrich eggs. The gaudy montbretia, which spreads where it wants and has no respect for civilized taste, is a good flower for him to have introduced.

MORNING GLORY

BOTANICAL NAME: *Ipomoea.*
FAMILY: *Convolvulaceae.*

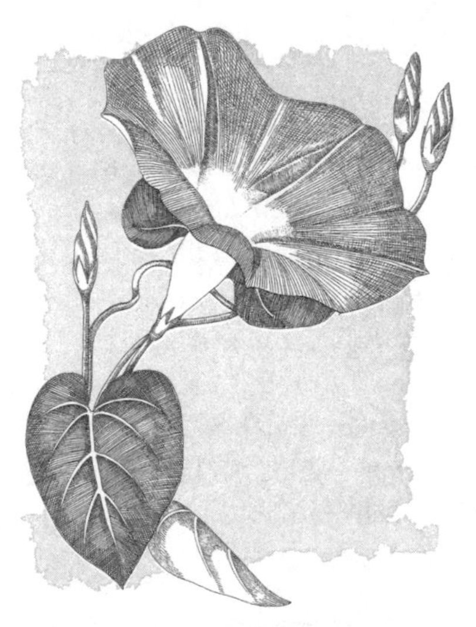

It is called "morning glory" because the flowers bloom early in the day and shut in the afternoon. If, as Reginald Arkell said, you don't look at them before breakfast, you probably don't see them at all. Its botanical name is from the Greek *ips*, "a worm," and *homoios*, "like," because of its worm-like stem.

Blue morning glory, called 'Heavenly Blue,' is native to tropical America and was introduced to Britain in about 1629. It was known as "Indian bindweed." According to whether you are a mouse or an eagle, looking at it from below or above, morning glory always twines clockwise or counterclockwise around its support. However you see it, the direction does not vary, regardless of heat, cold, light, climate, or even hemisphere. It is genetically programmed. The honeysuckle and the bindweed twist in opposite directions. Shakespeare may be referring to this when Titania tells Bottom that "I will wind thee in my arms, / So doth the Woodbine the sweet Honeysuckle / Gently entwist."

In Charles Darwin's study, *The Movements and Habits of Climbing Plants*, he describes experiments on the twining properties of plants, including the hop (which twists in the opposite direction to the morning glory). He rather touchingly includes in this scientific study that "my sons visited a hop-field for me" to confirm how the leaves were placed in relation to the spiral.

The bindweed or convolvulus, from the Latin *convolva*, "twisted," is a close relative of the morning glory, as is the sweet potato. Moonflowers are similar to morning glories, but their large white flowers open at noon. If both are grown on the same trellis, the moonflowers take over when the morning glories fade. They were also called *Calonyction*, from the Greek *kalos*, "beautiful," and *nyktos*, "night."

In *Old Herbaceous* (1951), by Reginald Arkell, Mrs. Charteris sees morning glories on the French Riviera, "as though someone had torn great masses out of a morning sky. It was so blue, so blue that it positively hurt." Her gardener obtains seeds from Kew and plants them in her greenhouse as a surprise:

> Once again she felt that tug at the heart, a kind of suffocation that almost hurt. Once again she was drifting over the blue sea, under a blue sky, into a lovely land of blue nothingness . . .
>
> "O, Pinnegar," she said, "how kind of you—how very, very kind!"

Anyone who has managed to get up before breakfast, to sit under a trellis loaded with blue morning glories in bloom, will know exactly how she felt.

MOUNTAIN LAUREL

BOTANICAL NAME: *Kalmia*. FAMILY: *Ericaceae*.

Linnaeus named the mountain laurel *Kalmia* after his pupil, Peter Kalm. In 1748, the Swedish government sent Kalm to explore Northern America and look for new plants, especially dye plants and medicines that might be useful to Sweden and would grow in northern latitudes. Kalm had accompanied Linnaeus to Russia and was loved by him "as was his own child."

Kalm's first description of the mountain laurel includes its possible uses. It was called the "spoon tree" by the Swedish settlers, he said, because "the Indians used to make their spoons and trowels of its wood." Its hard wood was also used to make the axles of pulleys and weavers' shuttles. The English called it a "laurel" tree because its leaves "resemble those of the Laurocerasus." Actually, even though it shares its thick, shiny leaves, mountain laurel is not even a member of the laurel, or *Lauraceae*, family, but is of the heath, or *Ericaceae*, family. The *Laurocerasus*, or

❧ Price drank the "laurel-water" and died in front of a delegation of fellows, headed by Sir Joseph Banks, that had come to talk to him about the fraud.

"cherry" or "English" laurel (known simply as "laurel" and widely grown there), is similarly named from the leaves and is a *Prunus*, or plum, a member of the *Rosaceae*, or rose, family.

Kalm's description of the mountain laurel includes details that not everyone would notice. Its poisonous leaves, when thrown on the fire, he said, "crackle like salt." If stags that have fed on it are shot and their offal is given to dogs, the dogs become very ill and "act as if drunk." This poison is called andromedotoxin and is shared by other members of the heath family (see "Rhododendron"). Leaves of the English laurel contain prussic acid, and an infusion of them was used by the botanist and fellow of the Royal Society James Price to commit suicide when his claim to change quicksilver into gold was disproved in 1782. He drank the "laurel-water" and died in front of a delegation of fellows, headed by Sir Joseph Banks, that had come to talk to him about the fraud.

The flowers of the mountain laurel have no scent but, said Kalm, "so equally and justly does nature distribute her gifts; no part of the creation has them all, each has its own, and none is absolutely without a share of them." He described the mountain laurels flowering in May, when "their beauty rivals that of most of the known trees in nature." The flowers are pink, fading to white, and "they resemble ancient cups." His journal is full of vivid little descriptions like this: how hummingbirds make sounds like "little turning spinning wheels"; how glowworms make the ground seem "as if it were sown with stars." His winsomeness shines through the journal. No wonder he was Linnaeus's favorite pupil. When Kalm returned to Sweden (accompanied by a young wife he met in America), Linnaeus rose from his sickbed to greet him. Kalm said that Linnaeus "because of the peculiar friendship and kindness with which he has always honored me has been pleased to call this tree . . . Kalmia latifolia." For once, however, one is inclined to think that it was the plant that was honored as well as the person. Peter Kalm was as distinctive and beautiful a person as the shrub he honors.

❧ Leaves of the English laurel contain prussic acid.

MYRTLE

COMMON NAMES: Myrtle, periwinkle. BOTANICAL NAME: *Vinca*. FAMILY: *Apocynaceae*.

What we call myrtle is really vinca or periwinkle. The name is used because the leaves are similar to the aromatic Mediterranean bush myrtle, or *Myrtus*.

Mediterranean myrtle was the symbol of both love and immortality, perhaps the two most important human preoccupations. It was used extensively and was considered an essential plant. Some stories said that Adam was allowed to take only three plants—wheat, dates, and myrtle—from Paradise. The myrtle nymphs taught the Greeks, via Apollo's son Aristaeus, the useful arts of making cheese, building beehives, and cultivating olives; they used myrtle to tan leather (which retained the aromatic fragrance) and as a black hair dye. The aromatic leaves are perforated (which later made it an important "Christ plant") supposedly because Phaedra, desperately in love with her stepson, Hippolytus, sat nervously pricking them while she watched him exercising. Unlike some Greek heroes, he behaved very properly and rejected her love, which turned to despair. It all ended in tragedy when she hanged herself (on a myrtle tree), and his chariot reins caught on some myrtle branches and he was dragged to his death.

The vinca, with its tough, shiny leaves similar to myrtle's (but not aromatic), probably came to Britain with the Romans. "Vinca" comes from the Latin *vincio*, to bind, and the long, tough stalks were used by the Romans to make ceremonial wreaths often associated with sacrifices. In the Middle Ages vinca wreaths garlanded criminals on their way to execution. "Vinca pervinca," or "creeping vinca," became "periwinkle."

The plant retained its associations with love. Albertus Magnus, a thirteenth-century Dominican botanist, scientist, and theologian who was canonized by the Catholic Church in 1931, said that "beate unto pouder with wormes of ye earth wrapped aboute it . . . [it] induceth love between man and wyfe if it bee used in their meals." Lovers prepared to go to those lengths might, one would think, be on the way to love anyway. Magnus added that if "said confection be put in the fyre," it turns blue—and maybe that is where the said confection often ended, while the man and wyfe managed without it. While not insuring, as far as we know, immortality, the Madagascar periwinkle is used to make vinblastine, which successfully treats Hodgkin's disease and childhood leukemia, thereby at least prolonging life.

NASTURTIUM

BOTANICAL NAME: *Tropaeolum.*
FAMILY: *Tropaeolaceae.*

Monet's famous garden at Giverny relied heavily on nasturtiums. They fitted the impressionist style of shimmering blurred colors, and they spilled over pathways like exuberant brush strokes. Monet was a contemporary of Gertrude Jekyll, and his garden, like hers, was revolutionary—a freely painted garden in an era of formal bedding. But as well as being informal and beautiful, nasturtiums are invaluable in the garden for filling in space with a minimum of effort and expense. As early as 1592 John Gerard observed that "one plant doth occupie a great circuit of ground."

Nasturtiums originated exclusively in South America, and were first described by the Spanish physician and plant collector Nicolas Monardes in *Joyfull Newes out of the Newe Founde Worlde* (1569). In the middle of every nasturtium petal, he noted, was a spot like "a droppe of bloode, so redde and so firmely kindled in couller, that it cannot bee more." But when most plants then were grown for usefulness rather than beauty, he said of the nasturtium, "I sowed a seede which thei brought me from Peru, more to see his fairnesse than for any Medicinall vertues that it hath." One can picture him at his desk with a nasturtium in front of him, gazing into the flower, careful not to miss the tiniest beautiful detail for his readers.

The name comes from the Latin *nasus*, "nose," and *tortus*, "twisted," because their pungent smell makes the nose wrinkle or twist. The botanical name is from the Greek *tropaion*, "a trophy," referring to the shield-like shape of the leaves. In ancient Greece, the shields and helmets of defeated enemies were fixed onto tree trunks. Linnaeus saw the plant twining up a post and thought the leaves looked like hanging shields and the flowers like helmets.

❧ A child can throw seeds in the ground and they will come up and cover the least fertile spaces with gorgeous shields and helmets.

Monardes's nasturtiums were *Tropaeolum minus*, smaller than the *Tropaeolum majus*, which came to Europe later, in 1648, and was the ancestor of our garden nasturtiums. Other nasturtiums include the Canary creeper, which is quite often grown in modern gardens as a summer vine, and a tuberous variety in Peru, used for food. Our garden nasturtiums are eaten too, and sometimes the seeds are pickled.

Thomas Jefferson planted nasturtiums every year. A letter from him to Bernard Peyton, not long before he died, said, "I missed raising Nasturtium seed the last year and it is not to be had in this neighborhood. Can your seedsmen furnish it?" He wanted enough seeds for a bed of nasturtiums ten by nineteen *yards*!

In the world exhibition in Paris in 1878, there were thirty varieties of nasturtium. They are so easy to grow that these days professional horticulturalists rather tend to ignore them. Of course there is not much point in their promoting them: a child can throw seeds in the ground and they will come up and cover the least fertile spaces with gorgeous shields and helmets. But their "fairnesse" is still irresistible, and no summer garden should ever be without them.

ORCHID

COMMON NAME: Orchid. BOTANICAL NAME: *Orchis*. FAMILY: *Orchidaceae*.

The history of orchids is of lust, greed, and wealth. The most famous orchid, the vanilla orchid, was thought to promote strength in the Aztecs, who drank vanilla mixed in chocolate. The name "vanilla" is derived from the Latin *vaina*, "sheath," and probably refers to the shape of the seed pod, or vanilla bean, but it has the same root as "vagina." At first Westerners did not appreciate it: "our Privateers . . . have often thrown [vanilla] away when they took any, wondering why the Spaniards should lay up Tobacco stems," wrote the pirate and botanist William Dampier, who was the darling of London society when not capturing ships and killing their crews. But by 1753, Linnaeus recommended vanilla as an aphrodisiac in *Materia Medica*, which listed sixty-nine species of orchid.

The name comes from the Greek *orchis*, "testicle." The tubers of Mediterranean orchids resemble paired testicles of different sizes, the smaller storing the previous year's food. The popular cattleya orchid was named in 1818 for William Cattley, who received it as packing around other plants. But after it flowered, it died, and wasn't found again for years. At a ball in Paris an orchid enthusiast noticed one in the cleavage of a South American ambassadoress. Immediately he inquired where it came from, and it was traced to Brazil. The cattleya lives up to the orchid's lascivious reputation in Marcel Proust's *Swann's Way*, when Swann offers to fasten one "a little more securely" in "the cleft of [Odette's] low-necked bodice." He then suggests he should "brush off" the pollen fallen from it, and the rest follows.

The cattleya lives up to the orchid's lascivious reputation in Marcel Proust's *Swann's Way*.

Other orchids are called "ladies' fingers" or "ladies' tresses," "long purples," or, as Ophelia says, "a grosser name." The paphiopedilum orchids are named for Phaphos, the site of a temple on Cyprus where Aphrodite was worshipped and prostitutes were available, and for *pedilon*, "a slipper." All in all orchids are, even in their names, closely connected with the power that "geveth lust unto the workes of generacyon and multiplycacyon of sperma" (Hieronymous Braunschweig, *Book of Distillation*).

The sexual behavior of orchids has baffled botanists since they first began to be studied. To germinate, their seeds need to be penetrated by fungus threads. Orchids go to extremes to propagate themselves, just as those who sought to acquire them went to extremes to show off their wealth and power.

In the nineteenth century orchids were collected by the ton. Once, four thousand trees were cut down for the orchids growing on them. One collector alone was said to have sent one hundred thousand orchids to England, many of which died. Wilhelm Micholitz sent home an orchid growing in a human skull, which was auctioned for a huge sum complete with container.

Orchid hunters mostly searched for riches rather than knowledge of the wondrous plant world. Nobody seemed to care that huge areas were stripped of native orchids, and we cannot much pity collectors who met with trouble. Even now, orchids are more often corsages for the rich than comfort for those who live in poor places. Their beauty, although undeniable, is not the beauty of simplicity.

• The vanilla orchid was thought to promote strength in the Aztecs.

OREGON GRAPE HOLLY

COMMON NAMES: Oregon grape holly, barberry, mahonia. BOTANICAL NAME: *Mahonia.* FAMILY: *Berberidaceae.*

Mahonia was brought from the far west by Lewis and Clark and called, by Thomas Nuttall, after Bernard M'Mahon, a refugee from political persecution in Ireland. M'Mahon found American gardening "in its infancy" and set to work to "introduce a love of flowers and fruit."

Within a few years, the catalogue from his seed shop in Philadelphia included one thousand species. The shop was presided over by his wife, and it soon became a meeting place for botanists. In 1806 he published *The American Gardener's Calendar*, which although not actually the first book about American gardening, was the first popular publication on this subject. It was an immediate success that for fifty years (during which time it was reprinted eleven times) was the standard American reference book on gardening. President Thomas Jefferson had a copy and bought seeds from M'Mahon. The correspondence between the president and the nurseryman continued until M'Mahon died in 1816. Seeds were ordered and sent, and there were friendly comments too. In April 1811, Jefferson told M'Mahon in a letter that "I have an extensive flower border, in which I am fond of placing *handsome* plants or *fragrant*. Those of mere curiosity I do not aim at, having too many other cares to bestow more than a moderate attention to them." In February 1812, M'Mahon wrote to Jefferson (and one can vividly picture the scene), "Excuse the confused manner in which I write, as there are several people in my store asking me questions every minute."

It was a temptation for American gardeners to make the New World an image of the Old. Only recently have we begun to appreciate that America does not have to mirror Europe. There were gardens here long before Europeans arrived, and some of our loveliest garden flowers are native to America. The fragrant flowers of the mahonia look like yellow lily-of-the-valley, the shiny holly-like leaves turn brilliant colors in autumn, and the blue-black fruit is edible and can make jelly or wine. It is a barberry and, strangely enough, related to the May apple. The

The fragrant flowers of the mahonia look like yellow lily-of-the-valley, the shiny holly-like leaves turn brilliant colors in autumn, and the blue-black fruit is edible and can make jelly or wine.

Latinized Arabic word *berberis* may have come via the medical school of Salerno, where it was an important medicinal plant.

But mahonia is its American name. That it bears M'Mahon's name is a tribute to a man who believed in American ideals and thought only of the "probable good I can render . . . to my fellow-men." M'Mahon said, "I do not begrudge a share to such of the brute animals as can possibly be benefitted thereby." This is the tribute of a man who maybe had universal ideals. But presumably it was not what he said when he found American groundhogs had eaten his garden.

OSWEGO TEA, BEE BALM, OR MONARDA

COMMON NAMES: Oswego tea, monarda, bee balm, bergamot. BOTANICAL NAME: *Monarda*. FAMILY: *Labiatae*.

Oswego tea is perhaps better known as "bee balm" or "monarda," but here it is called Oswego Tea in recognition that too few of our plants are known by Native American names, even though they were used for food or medicine long before Europeans came along. The Oswego Indians, who came from the region of the Oswego River (which means the "pouring-out place"), drank tea made from the monarda and taught the European settlers its uses.

Naming plants has always meant more than mere identification, and European settlers laid claim to American plants and animals by giving them European names. After the Declaration of Independence, Americans were eager to break all dominating ties with England, and we find new plants and animals being called for American botanists and explorers, even though the names were still Latinized. Native American names, however, were seldom used.

The Boston Tea Party led to a shortage of tea in America, so Oswego tea was used widely as a substitute for imported tea. Its scent is like that of oil of bergamot, one of the ingredients of Earl Grey tea. Earl Grey was on a diplomatic mission to China, and he had a special tea mixed for him with a secret recipe that he gave to Jackson of Piccadilly in 1830. Actually oil of bergamot comes from a kind of citrus called after Bergamo in Italy.

The name "bee balm" implies that the plant is attractive to bees. It is, but its long flower makes it less accessible to bees but easily accessible to hummingbirds. Hummingbirds were a source of delight and wonder to early American settlers. Peter Kalm (see "Mountain Laurel") wrote that "an inhabitant of the country is sure to have a number of these beautiful and agreeable little birds before his window all summer long, if he takes care to plant a bed with all sorts of fine flowers under them." This advice still applies today, for fashions in bird tastes alter less than fashions in gardening, and a bed of bee balm will be constantly visited by hummingbirds.

The name "monarda" is after Monardes (see "Nasturtium") and is appropriate because he was particularly interested in medicinal plants from the New World. As he said, "The corporalle healthe is more excellent, and necessarie then the temporall goodes," and he studied new plants, hoping to find new cures. The plant's botanical name is *Monarda fistulosa*, from the Latin *fistulosus* (hollow), because of the long pipe-shaped flowers. It doesn't really matter which of the four names you call it—as long as you plant it in your garden for the hummingbirds, because as Kalm further observed, "It is indeed a diverting spectacle to see these little active creatures flying about the flowers like bees."

PEONY

COMMON NAMES: Peony, paeony, pinny. CHINESE NAMES: *Chishaoyao* (red peony), *sho yu* (most beautiful). BOTANICAL NAME: *Paeonia*. FAMILY: Until recently *Ranunculaceae* (buttercup), now *Paeoniaceae*.

Pliny the Elder tells us that the peony received its Greek name from Paeon, the pre-Apollonian physician of the gods. In the *Iliad* there is a description of Paeon stanching wounds with herbs that thicken the flow of blood, like rennet curdling milk. Some stories say that the healing god Asclepias became dangerously jealous because Paeon possessed the healing root, and Zeus changed Paeon into a plant to save him. By the time of Pliny (who died in the eruption of Vesuvius, in A.D. 79) the peony was attributed with the power to cure twenty different ills.

Some of the plants that the Greek gods created to eternalize those they loved hardly seem worthwhile, because they aren't very long lived—but peonies can live for a hundred years or more if undisturbed. Indeed, they sometimes are the immortal remains of rural American families whose farms were abandoned and whose houses have crumbled. Where walls are barely traceable, a brilliant peony flowers in the wilderness of what was once a busy front yard.

How peonies came from China to Europe is too far back for us to know, but very probably they came to Britain with the Romans. The first peony was the *Paeony mascula* (or "male" peony), which was grown widely in medieval England, especially in monastery gardens. By 1548, *Paeony officinalis* ("medicinal" peony), our most common peony today, had been introduced from Crete. But the two kinds of peony were thought for some time to be masculine and feminine versions of the same plant. The *Paeony mascula* had a long, tapering root and pinnate leaves, and was used to treat male illnesses; the *Paeony officinalis*, which had feathery leaves, was used for female illnesses. John Gerard said that peony seeds glow in the dark, but he dismissed the belief that they could only safely be dug up at night and other "superstitious and wicked ceremonies . . . found in the books of the most Antient Writers . . . vainly feined and cogged in for ostentation sake."

John Gerard said that peony seeds glow in the dark, but he dismissed the belief that they could only safely be dug up at night and other "superstitious and wicked ceremonies."

Though ancient peonies probably came from China too, the first

"Chinese" peony, P. lactiflora (or "milky-flowered"), was sent to Joseph Banks at Kew in 1784 by the German naturalist Peter Pallas. It had been an important healing plant in China for centuries, but by this time the peonies' use in British gardens was mostly decorative, and when they were taken to America by settlers, it was not as healing plants.

Although their name comes from the healing powers of a physician, the word "paean" also means a hymn of praise—originally to Apollo. That perhaps is what we should associate with the name these days. With great puffs of exorbitant bloom they heal the spirit, if not the body, every spring; and once with us they are here forever. Could any flower more merit hymns of praise?

❧ Peonies can live for a hundred years or more if undisturbed.

PETUNIA

BOTANICAL NAME: *Petunia*. FAMILY: *Solanaceae*.

The petunia did not come to Europe from South America until the nineteenth century. At the same time it was being imported to France, Napoleon was putting his relatives on thrones all over Europe. His brother was on the Spanish throne, so there was no objection from the Spanish government when a French commission was sent to evaluate resources in South America and, in 1823, sent *Petunia nyctaginiflora*, or the "night-scented petunia" (now *P. axillaris*), to Paris. In 1831 James Tweedie sent *Petunia violacea*, or the "purple-flowering petunia," to the Glasgow botanical gardens. All our modern hybrids are descended from these two petunias.

Tweedie had been the head gardener at the Royal Botanic Garden at Edinburgh. But other passions pulled him, for when he was over fifty, he left this comfort and security and immigrated to South America. He supported himself by keeping a small shop in Buenos Aires, from where he went on botanizing trips all over the continent. Once he walked two thousand miles and returned so shabby and dirty his friends did not even recognize him.

> ❧ Once Tweedie walked two thousand miles and returned so shabby and dirty his friends did not even recognize him.

Unprotected Europeans out exploring were in danger of terrible, if justifiable, revenge if they ran into the persecuted native inhabitants. Tweedie survived these dangers and political intrigue and still managed to explore Patagonia when he was over seventy (once living on pinecones when he was near starvation). He died in Santa Catalina, aged eighty-six. In his life we see the pull of strange passions, and their rewards.

The young Charles Darwin, who was in South America at the same time, must have seen wild purple petunias growing everywhere, as Tweedie did, but there is no record of his sending them back to Joseph Hooker at Kew. By the time Darwin died though, in 1882, they were popular garden flowers. In 1834 John Loudon called petunias "the most splendid ornaments of the flower garden," and a Victorian gardener is said to have made a petunia bed twenty-one feet in circumference by training petunias over metal hoops to form a "table."

The name "petunia" comes from *petun*, a Brazilian word for

"tobacco," and petunias can be crossed with their tobacco cousins. Luther Burbank advertised a "nicotunia" plant, which was a cross between petunias and large-flowering nicotianas (see "Tobacco Plant").

◆ John Loudon called petunias "the most splendid ornaments of the flower garden."

Petunias continue to be hybridized to be stripy, fluffy, frilly, and generally as different as it is possible to make them. But the seeds of these hybrids will revert quickly to the small, aggressive purple wildflower that Tweedie and Darwin saw everywhere on their travels. We nurture the new hybrids, buy them afresh every year, and pluck off their sodden blossoms after every rain. If we wanted to, we could simply let the old purple petunias seed themselves every year, and they would pretty much fill our gardens, being strong enough to survive almost everything. But that, of course, probably won't happen—at least not while gardeners are still some of America's best shoppers and gardening is a multimillion–dollar business. So is tobacco, for that matter. Who would have thought that two South American weeds could have had such an influence on twentieth-century commerce and civilization? It only goes to show how little we know.

PHLOX

BOTANICAL NAME: *Phlox*. FAMILY: *Polemoniaceae*.

Native phloxes are to be found only in North America, but they belong to the widespread polemonium family, which includes the Jacob's ladder. The perennial phloxes from the East Coast came to Britain first. Their name was originally applied to another flower, described by Theophrastus but now unidentified, and is derived from the Greek *phlox*, meaning "flame." In 1732 a phlox was mentioned by the botanist Johann Jacob Dillenius as one of the many plants in Dr. James Sherard's famous garden at Eltham. Sherard commissioned Dillenius to write a description of his garden, and he also introduced him to Linnaeus, who, Dillenius had said disdainfully, had "thrown all botany into confusion." Linnaeus soon converted Dillenius so well that he was "in tears" when he left, and presented Linnaeus with a copy of his *Hortus Elthamensis* and several North American plants (which might well have included the phlox).

Sherard's phlox was the *Phlox paniculata*, which gets its name from its formation of flowers in panicles or loosely bunched clusters. It is the ancestor of our perennial border phloxes. Another eastern American phlox is *Phlox subulata*, named by Linnaeus from the Latin *subula* (an awl), referring to its pointed leaves. The name everybody associates with phlox, however, is that of Thomas Drummond, who sent the annual *Phlox drummondii* home to Britain.

Few gardeners can resist them, and Vita Sackville-West called them "monuments of solidity," but also said they smelled like pigsties.

Drummond was curator of the Belfast Botanic Garden and went to America in 1831 as an independent plant collector, exploring much of the Northwest by himself. He sent his guide away and spent one winter completely alone in a brush hut. He survived by chewing on an old deerskin when, because of snow blindness, he could not see to shoot game. He deterred grizzlies by rattling his specimen box at them, but this was less effective when he got between a mother and her cub and was nearly killed. He later survived a shipboard epidemic of cholera, nearly starved while wintering alone on Galveston Island, lost the use of his hand for two months, and suffered such severe boils that he was unable to lie down. In spite of all this, he applied for a grant of land in Texas, intending to bring his family over to America. In the meantime he went to Cuba, but in 1835, he died there from unrecorded causes.

One of the last plants he sent home was the *Phlox drummondii*, which Sir Joseph Hooker of Kew named in his honor, to "serve as a frequent memento of its unfortunate discoverer." Soon Victorian gardeners were developing and hybridizing it.

The phlox's beauty was appreciated from the start. Only seven years after Drummond's phlox was sent to Britain it was seen by James Drummond, Thomas's brother, growing in Australia. Peter Kalm included phloxes among the flowers that caused America to "abound with the finest red imaginable." They come in many colors but mostly variations of fiery red and magenta. Few gardeners can resist them, and Vita Sackville-West called them "monuments of solidity," but also said they smelled like pigsties. Whether this is true or not (and most people are unacquainted with pigsties these days), there are few modern gardens without them.

PLUME POPPY

COMMON NAMES: Plume poppy, tree celandine. BOTANICAL NAMES: *Macleaya cordata, Bocconia cordata.* FAMILY: *Papaveraceae.*

Plants of the beautiful plume poppy should be accepted from friends with caution. Gifts of them bring to mind the famous gardener, Miss Willmott, of whom Sir William Thistleton-Dyer said in her obituary: "As gardeners go she was not considered generous and one looked carefully at gift plants for fear they might be fearful spreaders." Though plume poppies spread rampantly, they are extraordinarily beautiful and on summer mornings the large leaves are covered with tiny dewdrops that catch the light against their dull blue-gray surface.

The hardy plume poppy is called *Macleaya cordata* or *Bocconia cordata*, from the Latin *cor*, "heart," because of its heart-shaped leaves, which really are more like the outstretched palms of blue hands, holding dozens of glinting pearls. The name "bocconia" is for the Sicilian botanist and medical doctor Paolo Bocconi. The plant first given his name would have been the *Bocconia frutescens*, which came from Peru or Mexico in 1739 and is not hardy.

The more robust Chinese plume poppy was sent to Kew Gardens by George Staunton in 1795. Staunton held the office of "minister plenipotentiary" on Lord George Macartney's unsuccessful expedition to China in 1792. This mission to obtain better trading conditions with China failed, despite presents for the emperor worth more than fifteen thousand pounds. But then the only person who spoke Chinese in the European party of one hundred strong was George Staunton's son, Thomas, who was eleven years old! Luckily for gardeners they managed to collect a surprising number of plants on the way back. The man who actually found the bocconia for Staunton was one of the two gardeners in the party, John Haxton. He was acknowledged botanically in 1831 when the haxtonia, a tree aster from Australia, was dedicated to him, but the name was later dropped.

The story of Macartney's embassy nevertheless became popular reading. In Jane Austen's *Mansfield Park*, written in 1814, the heroine, Fanny Price, is interrupted by her cousin Edmund, who resolves to save family propriety by taking part in a play he disapproves of. As he leaves to arrange this, he says to Fanny, "*You* in the meanwhile will be taking a trip into China, I suppose. How does Lord Macartney go on? (opening a volume on the table)." The plume poppy later acquired another name, *Macleaya cordata*, after Alexander Macleay, who was colonial secretary of New South Wales and secretary of the Linnaean Society in London.

The plume poppy has a plume-like flower, although it is usually grown for its distinctive leaves. As with all "invasive" plants, this characteristic can make it invaluable, if it can also be controlled. But that's a concept that doesn't only apply to the world of plants.

POINSETTIA

COMMON NAMES: Poinsettia, Christmas flower.
BOTANICAL NAME: *Euphorbia pulcherrima.*
FAMILY: *Euphorbiaceae.*

The poinsettia was named for Dr. Joel Roberts Poinsett, U.S. ambassador to the newly independent Republic of Mexico from 1825 to 1829. He was also a keen botanist and gardener, and he introduced the American elm to Mexico as well as sending the poinsettia to America. The poinsettia, whose color comes not from its flowers (which are an insignificant yellow), but from its brilliant bracts, was used in Mexico to decorate churches at Christmastime and called *flor de la noche buena*, or "Nativity flower."

The plant was not the only thing called after Poinsett. His policies in Mexico were unpopular, and the Mexicans coined the word *"poinsettismo"* to describe intrusive and officious behavior. He was an active politician at home and a member of the House of Representatives and of the Unionist party in South Carolina, which supported the Doctrine of Nullification—the rights of individual states to set aside federal laws that violated their "compact" with the American Constitution. In 1837 he was secretary of war. He was also a founder of the National Institute for Promotion of Science and Useful Arts, which later became the Smithsonian.

Though it comes from mostly tropical Mexico, the poinsettia is a short-day plant and only sets flowers when the nights are long and the days short. In its native country it grows to sixteen feet, but it is widely raised here in greenhouses for Christmas. If you wish to make it bloom again next Christmas, you must practice a little *poinsettismo* against its natural inclinations and cover it like a pet parrot early every evening so it gets no light.

It is a member of the euphorbia, or spurge, family, called after Euphorbus, physician to Juba, king of the ancient kingdom of Mauretania. King Juba was married to Cleopatra Selene, daughter of Antony and Cleopatra. There is a tradition that Dioscorides, who wrote the famous *De Materia Medica*, was Antony and Cleopatra's physician, so it is not too fanciful to suppose that Euphorbus had an interest in botany. The name "spurge" comes from the Old French *espurge*, and it was one of the powerful purgatives used in the Middle Ages to rid the body of "evil humors" like black bile and melancholy. Taken in quantity, however, the euphorbias are poisonous, and their sap can cause a blistery rash—so the Christ child's flower should be kept well away from animals and children.

❧ Mexicans coined the word "*poinsettismo*" to describe intrusive and officious behavior.

POPPY

COMMON NAMES: Poppy, popple *(Old English)*.
BOTANICAL NAME: *Papaver*. FAMILY: *Papaveraceae*.

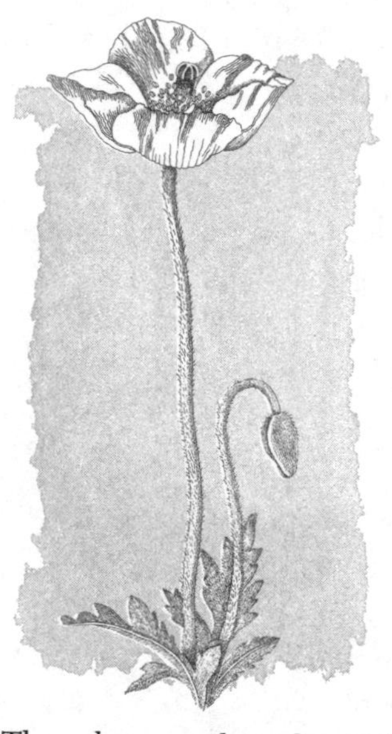

"An intensely simple, intensely floral flower," said John Ruskin of the field poppy in *Proserpina*. "All silk and flame: a scarlet cup . . . like a burning coal fallen from Heaven's altars. . . . No sparing of colour anywhere—no outside coarseness—no interior secrecy."

The botanical name is from the Latin *papaver*, possibly going back to *pap*, a milky food that could have associations with the opium poppy's milky juice. The field or corn poppy, *Papaver rhoeas*, takes its name from the Greek *rhoeas*, possibly from the root *rho*, "red." The substance rhoeadine is found in the flowers, and when you see the intense, glittering mass of a poppy field, it is not surprising to find that the color is from a specific chemical.

Corn poppies thrive in soil that has been freshly turned, because the seeds need light to germinate. This is the sad reason why they flourished so exceedingly in the battlefields of France during and after the First World War. The ground had been churned and turned by guns and battles, and poppy seeds, now exposed to the light, germinated by the millions. Forever after the red corn poppy was associated with war.

The dried petals of the corn poppy contain a soothing substance that the ancients used medicinally, but it is not comparable to the milky latex of the opium poppy, *Papaver somniferum*, which has been known medicinally since before we really have records. The opium poppy's name comes from the Latin *somnus*, "sleep." The narcotic opium is derived from this latex, which is allowed to dry on the slashed seed pods and then scraped off; it gets its name from the Greek *opion* or *opos*, meaning "vegetable juice."

The actual seeds of the opium poppy are not narcotic, and the dark seeds of the *Papaver somniferum* are used widely in bread and cakes. It is quite legal to buy the white seeds of *Papaver somniferum* var. *hortense* in America too, and in fact many nurseries sell them, but it isn't technically legal to grow poppies from them. The Oriental poppies, which are perennials, sound much more daring, but are in fact harmless as far as opium goes. Another garden poppy is the apricot and coral colored Iceland poppy, or *Papaver nudicaule* (from the Latin for "bare-stemmed").

Shirley poppies were bred from field pop-

It is quite legal to buy the white seeds of *Papaver somniferum* var. *hortense* in America too, and in fact many nurseries sell them, but it isn't technically legal to grow poppies from them.

pies by the Rev. William Wilks, vicar of Shirley, in Surrey. While admiring a patch of corn poppies in the field behind his house, he noticed one of them had a small band of white along the edges of its bright red petals. He climbed into the field (he was not, one imagines, wearing his cassock), marked the poppy, and later sowed its seeds in his garden. Every year he selected out all the flowers except his white-edged variety, which gradually changed until he had bred a rainbow of new colors on a pure white petal base.

Poppies are easy to grow but cannot be moved. The seeds have to be thrown down where they will be. Even then the dabs of impressionist brilliance they give our gardens are not always where we plan them. "The Poppy is the slyest magician of the whole garden," wrote Alice Morse Earle. "He comes and goes at will. This year a few blooms, nearly all in one corner; next year a blaze of color banded across the middle of the garden like the broad scarf of a court chamberlain. Then a single grand blossom quite alone in the pansy bed, while another pushes up between the tight close leaves of the box edging:—the poppy is *queer*."

PRIMROSE

BOTANICAL NAME: *Primula*. FAMILY: *Primulaceae*.

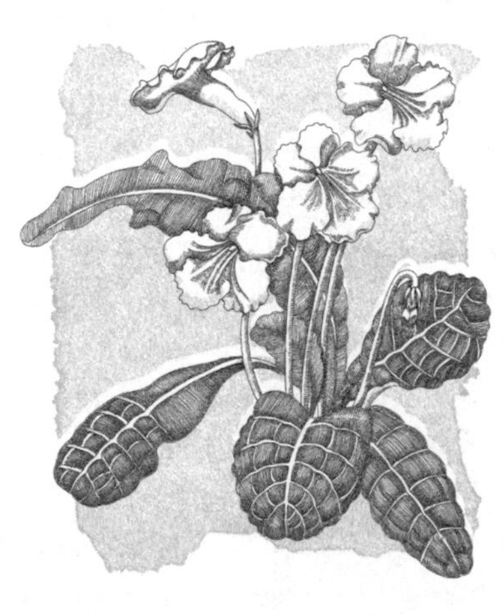

The primrose is the *prima rosa*, the "first rose" of the year. It was also named *primaverola*, from *fior di primavera*, the "first spring flower," or *primarole*, as in Chaucer's miller's wife who was a "primerole" and "blisful on to see."

The short-lived primrose years, the primose paths of dalliance—these were the seasons when our grandmothers warned us not to pick up young men on buses and we jumped on buses anyway and didn't care. The wild European primrose, *Primula vulgaris*, stood not only for spring but for first love, often half-hidden and growing suddenly in unexpected places, sometimes "no sooner blown than blasted." John Ruskin compares them to newborn yellow ducklings peeping out of their rosette of leaves.

Garden primulas, although of the same family, are different. They are sophisticates in the flower world, the darlings of breeders, often named for those who did not wish to be forgotten or half-hidden. Many came from China, where they had been hybridized for so long that the original wild ones were extinct. Others were collected in the Alps.

On November 21, 1861, Charles Darwin described the dimorphic condition of the primula to the Linnaean Society, commenting that plants with flowers showing the globular stigma at the mouth of the corolla are called "pin-eyed," and those displaying the stamens are called "thumb-eyed." This difference in structure favors cross-fertilization, which not only made breeding primulas easier, it also altered the appearance of the flower and so was important to breeders of show primulas, or auriculas (so called from the Latin *auricula*, "an ear," because the leaves are shaped like ears). Village children, Darwin said, noticed this difference in the flowers, as they could best make necklaces by threading and slipping the longer corollas of the long-styled flowers into one another.

❧ Willmott had a Napoleon fixation and built a hut in her garden which was an exact replica of one where Napoleon slept when crossing the Alps.

By the nineteenth century, florists' societies of specialized breeders, mostly working-class men, met all over England. There would be a dinner, with plenty to drink and speeches, before the flowers, displayed on elaborately and beautifully built wooden stages, were judged. The winners would be named for the breeders, conferring horticultural fame on them, or would be given names like 'Glory of England,' 'Privateer,' and 'Empress of Russia.' The rules about the flowers' appearance were rigid—they had to have a velvety texture and distinct hues, no shading, no pin-eyes, sometimes a gray or white edge, sometimes a "mealy" texture from a waxy powder on the petals. They were far removed from the modest spring primroses of hedgerows.

Miss Ellen Willmott, the famous rich spinster gardener who lived at the beginning of this century and had gardens in France, Italy, and England, was also an enthusiastic breeder of primula hybrids. A primula was one of the many flowers named *willmottia* after her. She herself was arrogant, extravagant, and, even by those who admired her, not really liked. She had a Napoleon fixation and built a hut in one of her gardens that was an exact replica of one where Napoleon slept when crossing the Alps. It was all a far cry from the little wild English primrose and what it stands for—the difference between carefree youth and the power of money. In the end, Miss Willmott lost all her money, and her great gardens were deserted. But perhaps wild primroses grew in her abandoned woods again and the tenderness of reckless first love triumphed after all.

❧ The wild European primrose stood not only for spring but for first love.

RED-HOT POKER

COMMON NAMES: Red-hot poker, torch lily.
BOTANICAL NAMES: *Kniphofia, Tritoma.*
FAMILY: *Liliaceae.*

Gardening is many things, amongst which is fashion, and gardeners are quick to dismiss as vulgarity what they do not embrace. Maybe because the flower stalks of kniphofia do look extraordinarily like actual red-hot pokers, they challenge our concept of a garden seeming to have been created by Nature rather than by ourselves—a concept that, ever since the time of Gertrude Jekyll, has tended to be the aim of fashionable gardeners.

The red-hot pokers came late to our gardens. The first were introduced from the Cape of Good Hope in about 1707, but they were thought to be greenhouse plants until the nineteenth century, when they were grown out of doors. According to Alice Coats they were particularly popular in the west of Scotland, where they were called "Baillie Jarvie's Poker." Baillie Jarvie, in Sir Walter Scott's *Rob Roy*, "seized on the red hot coulter of a plough . . . and brandished it with such effect that . . . he set the Highlander's plaid on fire."

Linnaeus first described the red-hot poker as *Aloe uvaria* because the flowers look like a little bunch of grapes (Latin, *uva*). But in 1794 the botanist Moench created the genus *Kniphofia*, in honor of Johann Hieronymus Kniphof, whose eighteenth-century *Herbarium vivum* was illustrated with "nature prints" made from actual inked plants. Meanwhile, another botanist, Ker Gawler, called them *Tritoma*. This name comes from the Greek *tries* (three) and *temno* (to cut) because of the three sharp edges on the ends of the leaves. They are still sometimes called "tritomas," even though the name "kniphofia" was settled on by botanists in 1843.

The *Kniphofia northiae* was introduced and named for Marianne North, a Victorian lady who traveled around the world by herself to observe and paint "tropical vegetation in all its natural abundant luxuriance." In her reminiscences she described cold, mud, hunger, and giant leeches sticking to her long skirts. She stayed for a while alone in a huge deserted mansion and delightfully described hanging as decoration a bunch of bananas where the chandelier had once been.

Kniphofias are beginning to be seen more often in American gardens. It sometimes happens with fashion that we look later with new eyes at what once we saw with scorn. The kniphofias, with their cumbersome name and amusingly pedestrian disguise, can suddenly be seen for what they are: rows of exquisite blossoms tiered to a point, shading gradually from fiery red to creamy white. They're not vulgar, they're beautiful.

RHODODENDRON

BOTANICAL NAME: *Rhododendron.*
FAMILY: *Ericaceae.*

The name "rhododendron" was originally applied to oleanders (and it still includes azaleas—see "Azalea"), but in 1563 Andrea Cesalpino mentioned the *Rhododendron ponticum* in *De Plantis*. It was introduced to Britain in the 1670s. The name "rhododendron" comes from the Greek *rhodon* (a rose) and *dendron* (a tree). The name *"ponticum"* comes from the territory of King Pontus, who was king of the sea and the son of Gaea, goddess of the earth. It was in the region south of the Black Sea where, aged fifty, the poet Ovid was banished and consoled himself by writing letters and poetry. Xenophon described Greek troops retreating from Asia Minor in 400 B.C. and being poisoned by honey from the *Rhododendron ponticum* after they raided some local beehives. The Romans would not accept the usual tribute of honey from Ponticum but took a double amount of beeswax from there instead. The nectar of members of the Ericaceae family, which includes the rhododendron and our mountain laurel, contains poisonous andromedotoxin, and beekeepers have to be careful where they put their hives in spring. Every year a few people get sick from toxic honey.

The Romans would not accept the usual tribute of honey from Ponticum but took a double amount of beeswax from there instead.

After the *Rhododendron ponticum*, the American rhododendrons were the next to come to Britain. The Quaker botanist John Bartram sent the wild *Rhododendron maximum* to Peter Collinson, who, on July 20, 1756, wrote that "this year the Great Chaemerhododendron flowered for the first time it is a Charming plant."

Other American rhododendrons and azaleas came to Europe, and "American gardens" containing them and American trees became fashionable. In the mid-nineteenth century Joseph Hooker transformed the landscape of Britain with rhododendrons sent from the Himalayas. His adventures included being captured for several weeks by hostile Tibetans and suffering such discomforts as leeches which, he said, "get into the hair and all over the body." Once he described having to sit on a plant until it thawed sufficiently for him to dig it out.

Rhododendrons continued to be brought from the Himalayas to Britain and one great collector, Lionel Rothschild, devoted his life and fortune to them, creating a magical park of them at Exbury (from which many get their name). Because, given an acid soil, rhododendrons are rewarding to grow, they are sometimes overdone. Purists dislike seeing them in gardens like Stourhead, which predate their introduction and do not fit the original design. Victorian-style shrubberies, impenetrable with rhododendrons, are gloomy and overwhelming and make one wonder if there is a murdered housemaid buried somewhere amongst them. Even so, they deserve our respect, if not our homage. When the *Rhododendron calophytum* first flowered at Wakehurst Place (now a branch of Kew), Lord Rothschild, Lord Aberconway, and Gerald Loder were seen in a procession, walking round and round the bush, raising their hats to it. Are we, perhaps, a little blasé?

ROSE

BOTANICAL NAME: *Rosa*. FAMILY: *Rosaceae*.

The rose represents love, magic, hope, and the mystery of life itself. Its name, ordinary enough, refers to its color (*rosa* is Latin for "red"). But that's like saying the heart is a muscle situated on the left side of the rib cage. The flower's mysterious associations date to the earliest civilizations—the Persian word for rose, *gul*, also meant "flower" and was close to *ghul*, the word for "spirit."

From earliest times the rose symbolized love and passion. The Greeks associated it with the blood of Aphrodite's beloved Adonis; the Romans used roses in feasts and orgies with such abandon that on one occasion the guests were actually smothered by rose petals falling from the ceiling. From an image of pagan love the rose was transfigured to an emblem of Christian mystical and spiritual love—connected with the Virgin Mary, with Christ's blood, and with the crown of thorns.

Before the sixteenth century there were a few basic roses in the West. The Apothecary rose, or *Rosa gallica*, native to Europe, was used by healers for almost any ailment, from barrenness (cured by swallowing a rose petal) to washing "molligrubs out of a moody brain." A striped Gallica called 'Rosa Mundi' commemorated Henry II's mistress, Rosamund, who was hidden by him in a labyrinth at Woodstock, near Oxford, but was tracked down by the jealous Queen Eleanor of Aquitaine and murdered. The Damask rose, which flowered twice a year, was thought to come from Damascus and was used to make rosewater. The Dog rose, or *Rosa canina*, was said to cure the bite of a mad dog.

> ❧ From an image of pagan love the rose was transfigured to an emblem of Christian mystical and spiritual love.

Hybrids and descendants of roses included the white *Rosa* × *alba*, which represented the House of York in the Wars of the Roses; the famous French 'Maiden's Blush' or *'Cuisse de Nymphe Emue'* (which is a seductive fleshy color); the spiny Eglantine (from the Latin *aculeatus*, "prickly"); the sweet-smelling Musk rose; and the Centifolia, or hundred-leaved rose, developed in Holland, with huge, sterile, cabbage-like flowers. During the sixteenth century the first yellow rose arrived from Persia.

At the end of the eighteenth century China roses came to Europe. These, unlike the old roses that at best only bloomed twice a season, bloomed continually. Among these were the Tea roses, which do not seem to have smelled of tea—some theorize that they were in the boxes along with imports of tea. They were tender plants that could not be grown out of doors until crossed with Hybrid Perpetuals, but afterward became the basis of nearly all our modern roses. The first pink Hybrid Tea rose, bred in France, was called 'La France' and was the rose most often given to our grandmothers by their "beaux."

All this breeding and rose history has left us many names of people and places, as well as their dreams. The Chinese, who have grown roses from prehistory, tended to name roses for poetic concepts, such as *'Yu-go-tain-tsing'* ('Clear Shining after Rain'). Many French roses, especially those bred during the time of the Empress Josephine, who had the most famous rose garden of all, were named after distinguished men, their wives, or their mistresses. Their identities often live on only in the roses that are named for them. Madame Testout, for example, was a dress designer in Paris; Madame Isaac Pereire was the wife of a French banker; Madame Hardy was the wife of the superintendent of the Luxembourg Gardens; Madame Plantier was the wife of Josephine's head gardener. All are now resurrected as roses. Josephine herself is perhaps more often mentioned as a rose than as Napoleon's sad ex-wife, who carried a rose with her always so she could hide her bad teeth behind it when she laughed. She caused ships carrying roses to be allowed through the battle lines unscathed by both French and English for the garden at Malmaison, also remembered in a rose.

Today's roses can tell us of the dreams and glamor of a bygone era, the magical people whom *everyone* knew (like Grace Darling, the brave coastguard's daughter), but they make us wonder, will our great-grandchildren ask who Grace Kelly or Princess Di were? Will Chrysler Imperial be as obscure as a spinning Jenny? Will Peace be a general hope or dream, rather than a commemoration of a rose bred in France and smuggled out to America just before the Nazis invaded? At least these names will be mouthed with delight each season, as the roses bloom again.

RUDBECKIA

COMMON NAMES: Coneflower, black-eyed Susan.
FAMILY: *Compositae.*

The best-known rudbeckia, a native of North America, is affectionately known as "black-eyed Susan," who figures in many ballads and songs. In the "Ballad of Black-Eyed Susan" by John Gay, she goes aboard a ship to ask the "jovial sailors" where her sweet William has gone. The plant's descriptive name, *hirta* (Latin for "hairy"), refers to its hairy stem.

Linnaeus called the coneflower *Rudbeckia* after Olof Rudbeck the Younger, who taught at Uppsala University and whose father had founded its botanical garden. In 1730 he offered Linnaeus a job tutoring his three youngest children.

Both Rudbeck and his father were leading scientists and botanists. Together they compiled a volume, called *Campus Elysii*, of all plants known at the time, illustrated with thousands of woodcuts. It was lost in a fire that destroyed much of the town of Uppsala in 1702. Rudbeck the Younger, still energetic although in his sixties, was working on a giant thesaurus of European and Asiatic languages when he met Linnaeus.

Linnaeus had just written a paper introducing his revolutionary theory on the sexuality of plants. His system had the beauty of simplicity. By counting male organs (stamens) and female organs (pistils), anyone who could count could sort plants into one of twenty-three classes. It became the most widely used system of classification until the early nineteenth century. Of course its blatant sexuality caused its own problems. Linnaeus referred to the stamens as "husbands" and the pistils as "wives," and the flower itself became the "marriage bed." Teaching botany had to be X rated, and by 1808 the bishop of Carlisle wrote despairingly that "nothing could equal the gross prurience of Linnaeus's mind" (see "Love-in-a-Mist").

Teaching botany had to be X rated, and by 1808 the bishop of Carlisle wrote despairingly that "nothing could equal the gross prurience of Linnaeus's mind."

Linnaeus had been so poor he used to block the holes in his shoes with paper and he was frequently short of food. But in Rudbeck's house, his days of poverty were over. He named the coneflower after his patron, saying, "So long as the earth shall survive and as each spring shall see it covered with flowers, the Rudbeckia will preserve your glorious name." He added that he had chosen a noble, tall plant that flowered freely and that "its rayed flowers will bear witness that you shone among savants like the sun among the stars."

There is another floral black-eyed Susan, the greenhouse vine *Thunbergia alata*, which was introduced from South Africa in 1772 by Thunberg (see "Japonica") and is often grown in America as a summer annual. It was named by Verduyn den Boer, who said, "As long as in our Paradise of flowers there wanders a single botanist, so long will the name of Thunberg be held in honoured remembrance."

Mostly neither botanist is remembered, and those black eyes of Susan have taken over. Even if we use their botanical names, we often do not remember whom they represent. But Rudbeck, who enjoyed three wives and fathered twenty-four children, seems, like Linnaeus, to have been no prude—and he is well commemorated by saucy Susan's flower.

SCARLET SAGE

COMMON NAMES: Scarlet sage, salvia. BOTANICAL NAME: *Salvia.* FAMILY: *Labiatae.*

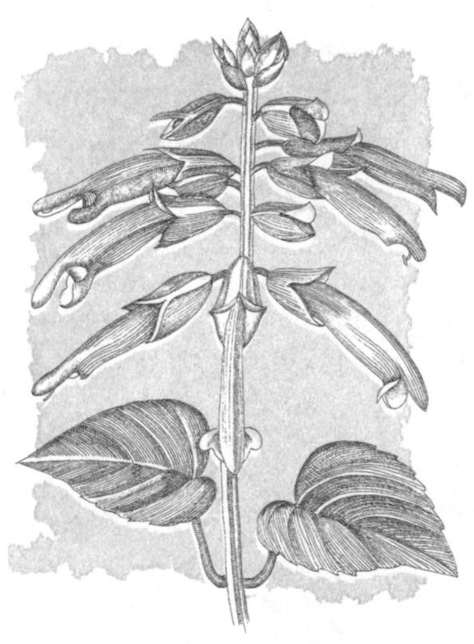

Scarlet sage is, like culinary sage, a member of the mint family. The name "sage" is the English corruption of *salvia*, derived from the Latin *salvus* (healed or saved). No garden of the past would have been complete without the medicinal or culinary sage, *Salvia officinalis*, but the scarlet sage is often rejected by gardeners nowadays as being too gaudy. Hummingbirds do not share this perception. When the scarlet sage was introduced to Britain in the 1820s, it immediately became a popular bedding "annual," although it is actually perennial in its native Brazil and Mexico. Most of the garden salvias grown today descend from it. Along with lobelia and ageratum, it could be part of a patriotic carpet bed—the kind of thing the Victorians loved. It is often still used in this way, thereby enhancing its reputation for vulgarity. But amongst other flowers in an informal bed, with hummingbirds darting in and out, its real beauty can be appreciated.

Baron Alexander von Humboldt, the German aristocrat who sent back the *Salvia splendens* from South America, wanted to go with Napoleon to Egypt in 1798 (see "Montbretia") but grew impatient waiting. So he went to Spain to organize and pay for his own voyage to South America. Humboldt had a theory that the immense vegetable richness of South America was to be explained by an unusually high level of magnetism in the area. "In the interior of this new continent," he said, "one almost grows accustomed to seeing man as not essential to the order of nature." This was a new concept and a new world that his writings opened up to contemporaries such as Charles Darwin. Plants, seeds, theories, and accounts of hair-raising adventures came back home. He experimented with curare, almost killing himself when a jar of it leaked over his stockings and was only discovered just before he put them on over feet raw with chigger bites. Another time, seven horses in their team were killed by electric eels when they were crossing the Orinoco River.

Humboldt and his fellow naturalist Aimé Bonpland returned in 1804 to a Paris where Napoleon Bonaparte had just become emperor and where the world's greatest scientists and philosophers were on hand to applaud their discoveries. Goethe said that "my natural history studies have been roused from their winter sleep" by Humboldt. Thomas Jefferson wrote, "You have wisely located yourself in the focus of the science of Europe. I am held by the cords of love to my family and country or I should certainly join you."

The scarlet sage, as brilliant as the man who brought it back with him, may be gaudy, and is too often used in a way that reflects the superficialities of our civilization, but the little hummingbirds that dart to it, back and forth, come each summer from a faraway world that still, as Humboldt said, has much to teach us.

SILVER BELL

COMMON NAMES: Silver bell, snowdrop tree.
BOTANICAL NAME: *Halesia*. FAMILY: *Styracaceae*.

The silver bell is a small flowering tree native to South Carolina. It was discovered and described by Mark Catesby, who explored the South from 1712 to 1719 and wrote a *Natural History of Carolina, Florida, and the Bahama Islands*. It is a beautiful little tree with white flowers hanging like tasseling bells along the underside of the branches, but it was not named for Catesby, who is immortalized instead in the bullfrog, *Rana catesbeiana*. The halesia is called after the Rev. Stephen Hales.

Hales was curate of Teddington and rector of two parishes, as well as being a fellow of the Royal Society and chaplain to the Prince of Wales. He helped Princess Augusta lay out the botanical gardens at Kew and designed the flues for heating the Great Stove, or greenhouse, there. In an alarming experiment, he made the first measurement of animal blood pressure. A white mare was tied to a gate, then thrown down, and blood was taken out of her neck through a goose's windpipe, joined to a glass tube where, as the reverend gentleman observed, it rose to a height of eight feet and three inches.

Taken in context, Hales's animal experiments were not as brutal as they seem to us now. At that time animals were not considered to have feelings. The circulation of the blood in humans and animals had been demonstrated by William Harvey in 1628, but until Stephen Hales's experiments, showing that root pressure, transpiration, and capillarity forced sap up plants, many thought that plant sap circulated like animal blood. Hales also concluded that "one great use of leaves is to perform in some measure the same office for the support of vegetable life, that the lungs of animals do for the support of animal life," but it was not until after his death that plant respiration could be properly studied. After Joseph Priestly discovered oxygen in 1774 (and Lavoisier named it), photosynthesis began to be understood. Hales, however, had done the groundwork. It is reassuring to know that he was also enthusiastic about his milder discovery that "an inverted tea-cup at the bottom of . . . pies and tarts [can] prevent the syrup from boiling over," and that he was seen just before he died painting "with his own hands the tops of the foot-path posts, that his neighbors might not be injured by running against them in the dark."

❧ Hales was an eccentric, with streaks of brilliance and a certain cold-bloodedness (by our standards at least).

The halesia tree is deliciously pretty and should be more widely grown. Bishop Henry Compton, who was Catesby's patron, called it the "Snowdrop Tree," and that name really suits it better than its botanical one. For botanical names do not always suit those they honor. Catesby seems to have been a simple character. He illustrated his *Natural History* himself and excused "some faults in perspective and other niceties" because "I was not born a painter." One can't see him sticking tubes into struggling mares. The snowdrop tree, with its beautiful flowers, would have suited him fine. Hales was an eccentric, with streaks of brilliance and a certain cold-bloodedness (by our standards at least). It might have been better if he had been honored with the name of the frog.

The halesia tree is deliciously pretty and should be more widely grown.

SNAPDRAGON

BOTANICAL NAME: *Antirrhinum.*
FAMILY: *Scrophulariaceae*

Snapdragons were favorite flowers in the earliest gardens, but the date of their introduction is not known. John Parkinson says they were found wild in Spain and Italy but in England they were "nourished with us in our Gardens," although "seldome or never used in Physicke by any in our days." They must have been grown then more for their beauty than their usefulness. John Gerard, however, quotes Dioscorides as saying that "the herbe being hanged about one preserveth a man from being bewitched, and that it maketh a man gracious in the sight of people." Snapdragon's botanical name is *Antirrhinum*, from the Greek *anti* (like) and *rhin* (a nose), referring to the snout-like shape of the flower. It was also called "calves' snout."

Gerard says that "the women have taken the name Snapdragon." Dragon plants might well have been named by women. Other dragon flowers are dead nettle (or Parkinson's *Lamium pannocicum*) and dragonhead (*Physostegia*), which we often grow in gardens and call "obedience plant." The interesting thing about these plants is that, unlike real nettles, they are *tamed* dragons, dead and stingless, and can be played with and will obey (the flowers of the obedience plant can be moved around the stem and will stay just where they are pushed).

Children will play with what is around, even in bombed-out buildings and stinking streets, if that is where they live. In rural Britain, before toys were cheap and widespread, children played with flowers, leaves, and sticks. Sometimes rural play was not so sweetly bucolic. They played with living creatures too, for they were taught, by the Church and everyone else, that nature's resources were the playthings of mankind. Boys had battles with plantain leaves and nuts, and girls made dolls with flower skirts (maybe boys did too, more discreetly). They made cowslip balls and daisy chains. For those too young or too gentle to capture frogs or other wild creatures, snapdragons made pets. As any child still lucky enough to have them in the garden knows, if you squeeze the sides of the flower gently, the dragon's mouth, complete with lashing tongue, will open and close.

Women, even of few resources, looked for ways to amuse children. Dragon flowers were fun, and they were also a way of taming and quieting dragons that very likely might still lurk in the wild and certainly walked in nightmares. Maybe they were vestigial memories of times when monsters really roamed the earth, "the inherited effects," as Charles Darwin said, "of real dangers and abject superstition during ancient times." Even today, when most children play with toys, not flowers, cute cartoon dinosaurs, sometimes stuffed and "huggable," are given to our children and convert the horrors of the past into acceptable history. If we wore snapdragons round our necks, we might feel even safer.

SPIREA

COMMON NAMES: Bridal wreath, spirea.
BOTANICAL NAME: *Spiraea*. FAMILY: *Rosaceae*.

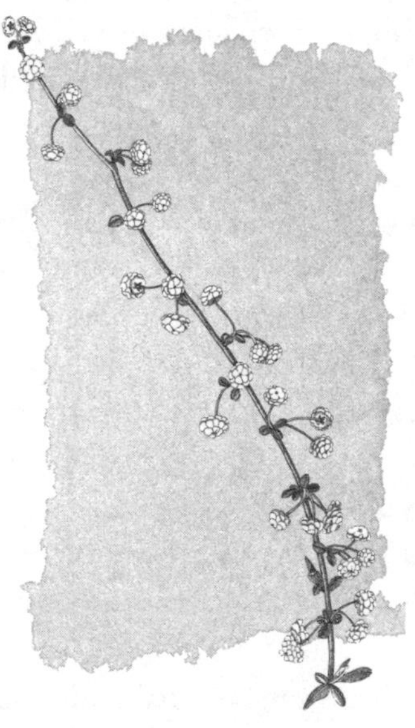

John Loudon recommended spirea for the "American gardens" popular in his time. These were gardens consisting of imported American trees and shrubs, and it was believed that a separate site was necessary for them because, as Loudon put it, they "do not harmonize remarkably well with European species."

Some nurseries, notably those of the Waterer family from Knaphill, Surrey, specialized in imported American plants. John Waterer had a nursery at Bagshot that was known as "The American Nursery," specializing in plants for "the Nobility, Gentry and others." The nursery continued in business until 1914, when it amalgamated with Bernard Crisp's nursery in Berkshire, and particularly specialized in rhododendrons and azaleas. One azalea, 'Pink Pearl,' was almost lost when stolen from the nursery but was found later in a cottage garden a mile away, demonstrating that even then (1897) theft was one of the hybridizers' problems. In our own more brutal era nurseries are often heavily guarded.

The spirea 'Anthony Waterer' came from Knaphill at the end of the nineteenth century, and it is still the most popular grown. It was bred from a Japanese spirea, introduced in 1870. It has reddish young foliage maturing to dark green, and the flowers are crimson-pink.

Spireas grow worldwide and were imported to Europe from America, Japan, China, and elsewhere in Asia. The Chinese reputedly used the flexible branches to make whips. Alice Coats observes that one Chinese name for spirea means "driving horse whip." They were well known in ancient Greece, where their whippy branches were used to make wreaths and garlands. The name "spirea" comes from the Greek *speiraiara*, which was a plant used in garlands, presumably named from the Greek word *speiros* (a spiral).

The Greeks had a pleasant interest in garlands and used them on sad, happy, and triumphant occasions. Dionysus is supposed to have made the first wreath out of ivy, and the use of wreaths spread to sacrificial animals, to priests, and to the people. In spring Athenians garlanded children who had passed the perilous period of infancy and reached their third year. Brides and grooms wore wreaths of flowers and heroes were crowned with them. Different flowers and evergreens were used for different occasions, depending on convention and availability.

Spireas may have been used particularly for weddings as many are covered with small white flowers that seem appropriate to brides. Anyway it was called "bridewort" early on in Britain and is called "bridal wreath" today. It is often interplanted with inflamed, red azaleas. Those who think this is a good combination might benefit from being bound with flexible branches and left to contemplate their creation for a while.

STOCK

COMMON NAMES: Stock, stock-gillyflower *(old)*.
BOTANICAL NAME: *Matthiola*. FAMILY: *Cruciferae*.

The herbalist John Gerard, after describing "Stocke Gillo-floures" growing in "most Gardens throughout England," concluded, tantalizingly, that "they are not used in Physick, except amongst certaine Empericks and Quacksalvers, about love and lust matters, which for modestie I omit."

A "gillyflower" was from *girofle* (see "Carnation"), a flower with the scent of cloves. A "stock-gillyflower" was one with a woody stock or stalk, like the cabbage, from the Anglo-Saxon *stocc* (a stick). Stocks came from the Mediterranean in the late Middle Ages or the early sixteenth century.

The botanical name, *Matthiola*, is after Pierandrea Mattioli, an Italian doctor who went to Prague as physician to the emperor Ferdinand I. He was most famous for translating Dioscorides's *De Materia Medica* into Italian and then Latin, along with his own commentaries. His work was so successful that it was reprinted repeatedly for the next two hundred years.

Stocks have been continually bred to produce new colors and better flowers. Weavers of Upper Saxony grew stocks competitively in their spare time and, in order to keep the shades distinct, only one color was permitted in each village. Philip Miller's *Gardener's Dictionary* (1731) describes "ten weeks" stocks, which "will produce Flowers in about ten Weeks after sowing." The Brompton stocks were bred in the famous Brompton Park Nursery, near London, whose value, according to Stephen Switzer in 1715, was "as much as all the Nurseries of France put together."

There were many theories on how to obtain the prized double stocks whose flowers look like small mauve and pink roses. As late as 1922, L. H. Bailey's *Standard Cyclopedia* said that the best way of obtaining doubles was to identify the seed pods (the doubles being shorter), after which the single ones could be removed "by hand . . . mostly by women and children." A Monsieur E. Chaté in

❧ "They are not used in Physick, except amongst certaine Empericks and Quacksalvers, about love and lust matters, which for modestie I omit."

—John Gerard

Traité des giroflées recommended taking off all the seed pods except a few, thereby increasing the sap flow to them and getting, he said, 80 percent of double flowers. The nicest suggestion, in an 1887 *Dictionary of Gardening*, is the method of "degustation of the buds": "the single plants can be recognised by their crispness and greater consistence, and can thus be weeded out. The disadvantage attending this method is that the plants . . . must all be grown up to the period when these buds are tolerably well advanced." Another disadvantage might have been the taste of the buds because, although said to be edible, stocks were only eaten in times of famine. But who knows; maybe there were compensatory effects. We were never told what happened to those Empericks and Quacksalvers whom Gerard described.

❧ Weavers of Upper Saxony grew stocks competitively in their spare time.

SUNFLOWER

BOTANICAL NAME: *Helianthus.* FAMILY: *Compositae.*

Helios was the Greek god of the sun. His flower is named from the Greek *helios*, "sun," and *anthos*, "flower," because the flowers always turn toward the sun. Helios was drowned by his uncles, the Titans, and then raised to the sky, where he became the sun. He himself was loved by a mortal, Clytie, who died of love for him, because he was indifferent to her—not surprisingly, because she had caused her own sister, whom Helios loved instead, to be buried alive. Clytie was rooted in her despair and followed the course of Helios's journey every day, just as does the sunflower, "who countest the steps of the Sun" (William Blake).

Sunflowers actually come from America, not Greece, so Clytie was probably turned into something other than helianthus. When they were first introduced, Nicolas Monardes described the flowers as "greater than a greate Platter or Dishe" and "marveilous faire in Gardines." John Gerard grew them, and winningly described the middle of the sunflowers "as it were of unshorn velvet, or some curious cloath wrought with the needle." The seeds, he said, were "set as though a cunning workman had of purpose placed them in very good order, much like the honycombs of Bees." They were reputed to grow to enormous heights, one report claiming they attained forty feet in the Padua botanic garden.

Crispian de Passe was an engraver who, in 1614, published *Hortus Floridus*. Until the invention of colored lithographs, illustrations were engraved and then painted by hand. The English version of *Hortus Floridus* includes detailed instructions about painting the sunflower: the petals were to be "tempered with a little lack, made shyninge and shadovved with sad yellow," the center was to be of a "berry yellow," merging into the "saddest of all." Early botanists and artists seemed almost to become a part of the flowers they described, making us feel their textures and follow the contours of their shaded colors.

The native American Jerusalem artichoke is also a sunflower, called "Jerusalem" as a corruption of the Italian *girasole*, "turning with the sun." It is edible but contains inulin, which is indigestible and causes "a filthie loathsome stinking winde within the body" (John Goodyer). Sunflowers were grown more for their usefulness than their beauty. The oil from their seeds is used for food, soap, paints, and cosmetics, among other things. They are high in protein and minerals. In North America they are grown and sold widely as bird food, because our destruction of natural habitats means that birds, if we want them in our gardens, have become increasingly dependent on us for their food, which we serve to them on "rustic" bird tables. This seems an inside-out procedure—although not from the point of view of a capitalist economy. The birds don't care either way, as long as they can find food and shelter—and they might be lucky and find a rustic birdhouse too.

SWEET PEA

BOTANICAL NAME: *Lathyrus odoratus.*
FAMILY: *Leguminosae.*

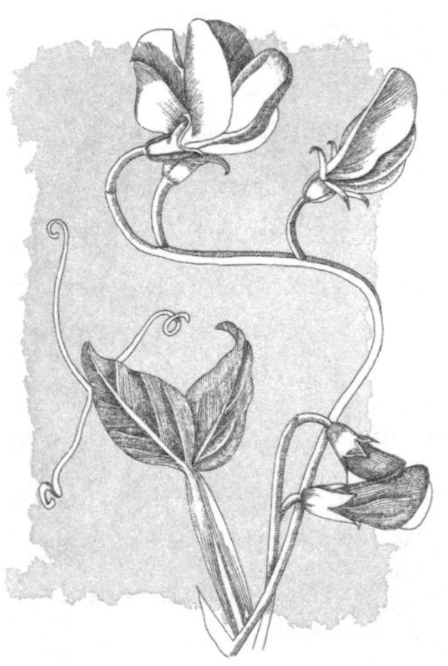

Sweet peas are, not surprisingly, members of the pea family; their botanical name is Greek for "pea." Sweet peas are *Lathyrus odoratus*—or "fragrant peas." But although they look alike, there is an important difference. Most peas are edible, including the wild, or "sea" peas, which in 1555, a year of famine, "miraculously" appeared on beaches of the Suffolk coast and saved the lives of the starving poor who gathered them by the sackful. But sweet peas are poisonous, and there is even a medical term, lathyrism, to describe sweet pea poisoning, which has serious consequences including convulsions, paralysis in the legs, and unconsciousness.

Sweet peas, so called for their sweet scent, are latecomers to our gardens. They were discovered by a Franciscan monk in Sicily, Father Franciscus Cupani, who wrote a description of them in *Hortus Catholicus*, which was published in 1697. In 1699 he sent seeds to Dr. Robert Uvedale, who was headmaster of Endfield Grammar School. Dr. Uvedale was a "very methodical and curious" botanist who was especially interested in hothouse plants or "rare exotics." He was one of the earliest hothouse owners in Britain and had six or seven of them. He was also the happy owner of a myrtle tree "cut in the shape of a chaire." He first raised sweet peas in his hothouses but then found them to be hardy outside.

The original blossoms, though fragrant, were small and purple. They were painted by Redouté, and Thomas Jefferson planted sweet peas in an oval bed accompanied by an exotic unknown called *Ximensia encelioides*, sent him by André Thouin in Paris. Still, the flowers did not become popular until the mid-nineteenth century, when they were "improved." From the 5 varieties originally available, there were 264 varieties shown in the famous Crystal Palace. The first frilled variety was bred by Silas Cole, gardener to the earl of Spencer, and it was tactfully named 'Countess Spencer.'

As the flowers grew larger, their scent was sometimes neglected in breeding. With the sudden influx of tropical, often scentless, plants into their gardens, Victorian horticulturalists were often more preoccupied, as indeed we have been, with splashes of color than subtle scents. The romantics still talked about sweetly scented cottage gardens and even the medicinal benefits of flower scents—Oliver Twist recovered his health in a garden of flowers that "perfumed the air with delicious odours." But fashionable flowers got bigger and brighter and often less fragrant at the same time. Sweet peas became larger and frillier and began to be bred in the colors of butcher's offal. Then there was a reaction. Cottage gardens and sweetly scented flowers began to be treasured, and earlier varieties of sweet peas were revived. Gardening, like all fashions, is apt to come round full circle.

TOBACCO PLANT

COMMON NAMES: Tobacco plant, nicotiana, flowering tobacco. BOTANICAL NAME: *Nicotiana alata*. FAMILY: *Solanaceae*.

Jean Nicot, who was the French consul in Lisbon in 1560, "wente one daie to see the Prysons of the kyng of Portugall" and was given a tobacco plant by the keeper of the prisons, who had obtained it "as a strange Plant brought from Florida." Nicot planted it in his garden where "it grewe and multiplied marveilously," and used it to cure a young man of an ulcer in his nose. He then applied it to the ambassador's cook, who had "almoste cut of his Thombe, with a greate Choppyng knife," and "from that tyme fourth the fame of that same hearbe encreased in suche sorte, that manye came from all places to have that same hearbe."

The garden tobacco plant, *Nicotiana alata*, is a close relative of the *Nicotiana tabacum*, or smoking tobacco. It is called *alata* (Latin for "winged") from the way the petioles, or leaf stalks, are set onto the plant. When tobacco was introduced to Europe, it was brought "to adornate Gardeines with the fairnes therof, and to geve a pleasunt sight," and then acclaimed as a medicinal plant and grown "more for his vertues, than for his fairenes." Nicolas Monardes named it after Nicot, "to the ende that he may have the honour thereof, accordyng to his desert." He devoted fifteen pages of his book about New World plants, *Joyfull Newes out of the Newe Founde Worlde* (see "Nasturtium"), to its uses and virtues.

We grow nicotiana in our gardens mainly because of its wonderful fragrance, especially in the evening, when the huge white flowers open to attract night pollinators. Anyone making an evening garden will want to include it. It is actually a perennial, but it is usually planted indoors early and treated as an annual, although Victorians sometimes kept it as a perennial houseplant.

The tobacco family hybridizes with ease, and presumably one could smoke petunia leaves or use the night-scented tobacco plant to make the healing ointment that John Gerard recommended to give to "thy wounded poore neighbor." Linnaeus, who was a heavy smoker, recommended smoking as a protection against infection. Infused tobacco leaves were used until well into this century as an effective insecticide, and pure nicotine is one of the most powerful plant poisons in existence.

We are still arguing, as they did in the sixteenth century, about how or if tobacco should be used, but no one argues about the beauty of the garden tobacco plant. Victorian gardeners recommended that it should be planted along paths, so at dusk they could walk outside and enjoy its heavy fragrance. The moth, fluttering toward the fragrant white flower funnels, glowing out of the dark border, is about to fulfill its own and the plant's destiny, which has nothing to do with us. We can go on arguing—enjoy or abstain; the moth will continue to flutter, and the flower to bloom, regardless.

TRUMPET VINE

BOTANICAL NAME: *Campsis* (formerly *Bignonia*).
FAMILY: *Bignoniaceae*.

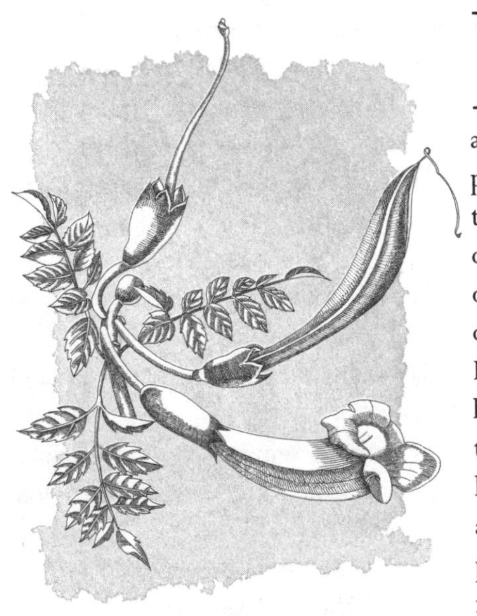

Doubtless few of us give much thought to Louis XIV; we are more apt to be preoccupied with love and war, taxes, politics, or even gardening. If, however, we find ourselves occasionally pondering the era of the Sun King, we probably think of his *affaires de coeur*, his military campaigns, his unpopular means of raising money, and the intrigues of his stupendous court. Gardeners might also think of Versailles, a "garden" that was really a symbol of power and prestige and unlimited control over nature. What we are not apt to think of is plants, or of the age of Louis XIV as a time of botanical discovery—either theoretical or actual.

In fact, the Jardin des Plantes (or King's Garden) and the University at Montpellier were expanded dramatically under Louis. Exotic new plants were raised in greenhouses, systems of classification were explored, and botanists, adventurers, and priests were sent worldwide to bring back discoveries. Plants represented new knowledge, and new knowledge represented power, power being what Louis wanted. The king's missionary priests, tactful about strictness of conversion, were accepted in China, and a wealth of new plants was sent home, including asters, delphiniums, jasmine, tea, soybeans, ailanthus, hibiscus, mulberry, peonies, and thuja.

❧ Exotic new plants were raised in greenhouses, systems of classification were explored, and botanists, adventurers, and priests were sent worldwide to bring back discoveries.

The Abbé Jean Paul Bignon, for whom the trumpet vine was first named, was Louis XIV's librarian. Joseph de Tournefort (see "Bear's Breeches") named the bignonia to express his "Esteem and Veneration" for Bignon, who had successfully nominated him to the Royal Academy of Science.

The American trumpet vine, *Bignonia radicans*, was probably introduced by the elder John Tradescant (see "Aster"). It was not grown much until after the Chinese trumpet vine had been introduced (in the eighteenth century) and crossed with it by the brothers Tagliabue, nurserymen from Laniate near Milan. In about 1889 the red-flowered 'Madame Galen' appeared. It is still our most popular garden trumpet vine.

The bignonia family was later divided by botanists, and the trumpet vine was no longer called after Abbé Bignon, but given its own classification as a campsis. This comes from the Greek *kampe* (bent) and refers to climbing tendrils, which had been described by Tournefort as "curls or tufts." Today's bignonia vine is the "crossvine," also a climber, with peach-apricot flowers. Its wood when cut transversely is marked with a cross. Other bignonias are greenhouse climbers, and there still seems to be some confusion about the name. The family though, includes the common catalpa tree, named *Bignonia catalpa* by Linnaeus to include the Indian name of *"catalpa."* In 1788, the *Flora Caroliniana* of Thomas Walter called the American species *Catalpa bignonioides*, which means "bignonia-like," and so it remained.

There isn't, to the common eye, much similarity between the trumpet vine and the catalpa—except that both are beautiful, both are American, and both should be in our gardens. Botanists live in their own world, however, and it's a useful one. Who are we to argue with them? After all, we have plenty to think about, like love and war, taxes, politics, and gardening.

TULIP

BOTANICAL NAME: *Tulipa*. FAMILY: *Liliaceae*.

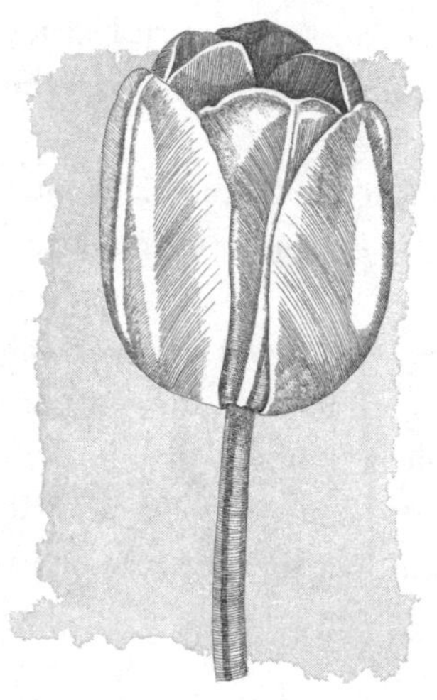

Tulips were probably introduced to Europe in the sixteenth century by Ogier Ghiselin de Busbecq, who was ambassador of the Holy Roman Emperor, Ferdinand I, to the court of Suleiman the Magnificent, sultan of Turkey. The name "tulip" is generally thought to be the Latinized name of the arabic *dulband*, "turban." It seems that Busbecq mistook the name of the flower for that of the turbans into which tulips were commonly tucked.

Once introduced, tulips rapidly became very popular. In 1593 Charles de l'Ecluse, or Clusius, grew tulips in Leyden. Clusius was director of the emperor's garden in Vienna and author of *Rariorum Plantarum Historia*. Clusius apparently asked such a high price for his bulbs that no one could buy them, and instead they were stolen from his garden. They soon spread throughout the seventeen provinces. Clusius visited England and saw Drake and Raleigh; maybe he even took bulbs with him because, by 1629, John Parkinson called tulips "the delights of leasure." He added, chauvinistically, that although to cultivate them women "are herin predominant, yet cannot they be barred from your beloved, who I doubt not, will share with you in the delights as much as is fit."

By the eighteenth century, the names of tulips included 'Semper Augustus,' 'Alexander the Great,' 'Artaxerxes,' 'Black Prince,' 'Duke of Vendôme,' 'Emperor of Germany,' and 'Duke of Marlborough.' In Joseph Addison's famous satire in *The Tatler* in 1710, he pretends he has overheard a conversation about these men, who turn out to be tulips, "to which the gardeners, according to their usual custom, had given such high titles and appellations of honour." The crowning point of the satire, however, is that he could show his friends "a chimney-sweeper and a painted lady in the same bed, which he was sure would very much please them."

Tulips, first observed by Westerners in men's turbans, seem to be a somewhat male-dominated flower. They were used as units of speculation during the famous "tulipomania" in Holland, when the numbers financially represented by them were far greater than the actual bulbs grown. Various disaster stories were told of bulbs being eaten and great fortunes being lost, and of their being exchanged for vast sums of money, houses, and carriages. One account tells of a breeder who gave his blanket to a tulip bed and died of cold. All these adventures involved men who wanted to make money. The most prized and expensive tulips were the "broken" or striped flowers. One reason these were good speculations was that it was not known, until recently, how the "breaking" occurred—it could just appear, and

❧ One account tells of a breeder who gave his blanket to a tulip bed and died of cold.

make the grower rich. Now we believe it is usually the effect of a virus, transferred by aphids. In 1637 the tulip market in Holland crashed, and finally the government forbade speculation in them. But by then the Dutch had become skilled in growing tulips, and the plants gradually became commercially important in Holland again.

In 1637 the tulip market in Holland crashed, and finally the government forbade speculation in them.

Turks, Pierre Bélon (see "Lilac") said, wore tulips "snugly in the folds of their turbans [but thought] little of their smell. They are unscented flowers we do not associate with solace and romance, but once they must have grown wild on Turkish hillsides, unexpected treats of beauty that would surprise whomever came across them. The tender, isolated blossoms, piercing the scrubby slopes, must have been loveable indeed, before they were transformed by the rich, the powerful, and the greedy.

VIOLET AND PANSY

COMMON NAMES: Violet, pansy, heart's-ease, Johnny-jump-up, love in idleness. BOTANICAL NAME: *Viola*. FAMILY: *Violaceae*.

According to John Gerard, the name "viola" came from Io, "the yoong Damsell" whom Zeus loved but changed into a heifer to protect her from his jealous wife, Hera. Zeus gave Io a field of violets to eat, "which being made for hir received the name from hir." Hera saw the tender little white heifer with purple violets in her mouth, and such perfect beauty aroused her suspicions. She asked Zeus to give her the calf, and he was trapped into assenting. Hera then had Io in her spiteful power and harassed her mercilessly, finally sending a gadfly to torture her until, unable to sleep or eat, she plunged madly into the Ionian Sea, also named after her. Finally Hera was able to extricate a promise from Zeus not to look at Io again, and, in exchange, turned her back into a girl.

Violets, then and later, were linked with love. The fairy spirit Puck employed their juice in *A Midsummer Night's Dream* to make "man or woman madly dote / Upon the next live creature that it sees." Most

Elizabethans called them "heart's-ease" and often associated them with an innocent, unspoilt love; they are appropriate for that because the violet flowers do not produce seeds, "as in all other plants that I know" (Parkinson). The seeds come from unopened, self-pollinating flowers later in the year, a quality called "cleistogamy."

In 1810 Thompson saw the first "blotched" pansy, "a miniature impression of a cat's face steadfastly gazing at me," as a stray flower in a bed.

When Napoleon was banished to Elba, he said he would "return with the violets." When he did return, Josephine was dead, and he picked violets from her grave before being exiled again to St. Helena. They were found in a locket, along with a lock of her hair, when he died.

The tricolored violet, or Johnny-jump-up, was the ancestor of our pansy, which is a relatively new flower, bred by Mr. T. Thompson, gardener to Lord Gambier who was Admiral of the Baltic fleet. Thompson crossed varieties of *Viola tricolor* with the yellow *Viola lutea* and *Viola altaica* (from Turkey and the Crimea). In 1810 Thompson saw the first "blotched" pansy, "a miniature impression of a cat's face steadfastly gazing at me," as a stray flower in a bed. Afterward, pansies became more and more "fancy" and were a popular florists' flower, grown and shown competitively. Purple violets were still used as boutonnieres in "pink" hunting jackets, and many were grown in conservatories for this purpose.

In the nineteenth century huge quantities of violets were also grown for perfume, especially on the French Riviera. By 1893, however, two German scientists, Tiemann and Kruger, had discovered the chemical formula of violet scent (which is also present in orris root—see "Iris") and patented it, calling their product "Ionone."

A pansy, or *pensée* (from the French *penser*, "to think"), is what is in our thoughts, and we rely on purity of thought not to see the world crooked. Violets and pansies represent love, but love in its highest form. Violets, as John Gerard said, "do bring to a liberall and gentle manly minde, the remembrance of honestie, comlinesse, and al kindes of vertues." "God send thee Heartsease," wrote William Bullein in 1562, "for it is much better with poverty to have the same, than to be a kynge with a miserable mind. Pray God give thee but one handful of Heavenly Heartsease which passeth all the pleasant flowers that grow in this worlde."

❧ The tricolored violet, or Johnny-jump-up, was the ancestor of our pansy.

WATER LILY

BOTANICAL NAME: *Nymphaea.*
FAMILY: *Nymphaeceae.*

Monet's famous water garden may have been inspired by M. Bory Latour-Marliac's new water lily hybrids, which were just becoming available in France. In 1898, Latour-Marliac told the Royal Horticultural Society how he had made these hybrids, crossing tropical water lilies with European varieties by a complicated process: the stamens have to be cut "at the very first moment of expansion" and the stigma brushed with the pollen of the crosses; after fertilization, the ovary sinks and ripening takes place underwater; the ripened seeds float up, looking like small pearls, and have to be collected at once or they soon sink again.

In warm climates, though, the most spectacular water lily is the gigantic *Victoria amazonica*. The explorer Aimé Bonpland is said to have tumbled into the water with astonishment when he first saw it. The Amazon water lily first flowered in England in 1849, in a tank built especially for it by Joseph Paxton, gardener to the duke of Devonshire and designer of the Crystal Palace (which he modeled on the structure of the amazonica's ribbed leaves). Queen Victoria was presented with a flower and told the plant was to be named *Victoria regina*, after her. But when it was found that in 1832 the explorer Eduard Poeppig had already described it and named it *Euryale amazonica*, after one of the three Gorgons, the rules of nomenclature would not bend, and no one dared tell the queen its true name. Then botanists decided that the amazonica was not of the euryale family, so it could be renamed *Victoria amazonica*. This didn't help matters though, because any connection with Amazons was felt to be "totally unsuited . . . with the name of Her Most Gracious Majesty." So the proper botanical name was simply kept a secret until after Queen Victoria had died, and the amazonica could be properly labeled at Kew.

The rules of nomenclature would not bend, and no one dared tell the queen.

The Egyptians revered the water lilies and lotus flowers, emerging from the water in the morning and sinking again when the sun set, like the passing of the day. Statues of the god Osiris, who was murdered by his brother Set, were bedecked with them. Set chopped up and scattered the dead body, but Osiris's wife/sister, Isis, collected all the pieces except, some accounts say, the phallus, which was eaten by a crab. Anyway, Isis was able to revive Osiris sufficiently so that (with or without the crab's dinner) he could conceive the solar god, Horus.

Water lilies are said to be edible and their root, according to Herodotus, tastes "fairly sweet." The botanical name is from the Greek *nympha*, "a water nymph and a virgin." The ancient Greeks believed they had antiaphrodisiac properties, and in the Middle Ages nuns and monks made "electuaries," or pastes, of ground water lilies and honey to preserve chastity (but we do not have the instructions on how to apply these).

Water lilies are not used as food now, or probably as antiaphrodisiacs either—the latter being a medicine we don't seem to require much in our present society. But the miracle of the pristine flowers, growing from murky waters in muddy ponds, is as powerful as it ever was.

WEIGELA

COMMON NAMES: Weigela, weigelia. CHINESE NAME: *Noak chok wha.* BOTANICAL NAME: *Weigela* (formerly *Diervilla*). FAMILY: *Caprifoliaceae.*

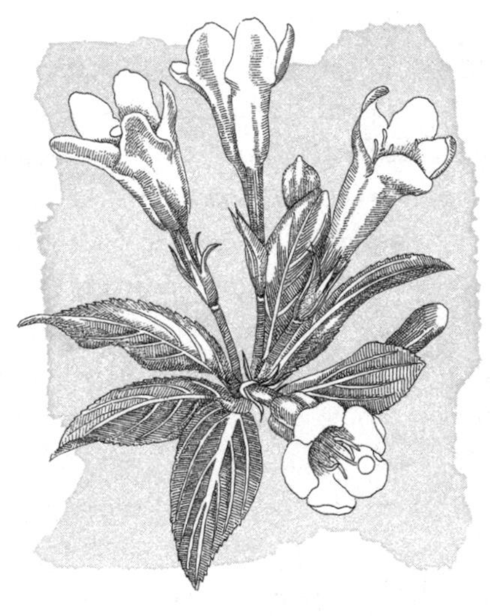

Our Oriental weigelas had a name change. They were once thought to be related to the *Diervilla lonicera*, or bush honeysuckle. This plant interests botanists because after the yellow flowers have been fertilized they change color to a deeper yellow, "to let the bees know the larder is empty," Neltje Blanchan wrote. The original "diervilla" was brought from Canada to France by a French surgeon, M. Dierville, who sent plant specimens home from French Acadia (Nova Scotia). Joseph de Tournefort (see "Bear's Breeches") named the bush honeysuckle in his honor. In 1708 Dierville published an account of his voyage in verse. The diervilla is not grown much as a garden plant but is widespread in northern America.

Robert Fortune found what was first known as the *Diervilla florida* in China in 1845, and he wrote a vivid description of his discovery. It was on the island of Chusan, in a mandarin's garden that he called "an excellent specimen of the peculiar style so much admired by the Chinese in the north . . . and generally called the Grotto, on account of the pretty rockwork with which it was ornamented." The new shrub grew only in the north of China and was not found wild—in fact it probably came from Japan to China originally. Fortune saw the plant covered in "fine rose-colored flowers, which hung in graceful bunches from the axils of the leaves and the ends of the branches," and he "marked it as one of the finest plants of Northern China." He said, "Every one saw and admired" it, and it was "a great favourite with the old gentleman to whom the place belonged." It is rather nice to think of the old Chinese gentleman enjoying his bush, all those years ago.

It is rather nice to think of the old Chinese gentleman enjoying his bush, all those years ago.

Fortune immediately sent specimens of it, together with a drawing, to the Royal Horticultural Society. It was soon established in British gardens and was said to be one of Queen Victoria's favorite plants. Other diervillas followed from Japan and Korea, and were hybridized with enthusiasm.

The Asiatic diervillas were thought to be a new species and were renamed *Weigela* after Christian Ehrenfried von Weigel. Weigel was a professor at Greifswald in northern Germany and the author of a floral tome called *Flora Pomerano-Rugica*. Botanists never really settled on the correct name of the weigela but only a few pedantic (and probably English) nurseries still call it "diervilla." Both it and the bush honeysuckle, which *is* undisputedly diervilla, are members of the caprifolia family, charmingly so named from the Latin *caper* (a goat) and *folium* (a leaf)—because the leaves caper all over the place, like little goats.

Immortalization is often unpredictable. Dierville and his descriptive verse have mostly been forgotten. Weigel's life's work must hardly be read, but everyone knows the weigela, and it is in most gardens. One wonders if Weigel would have been pleased by this, or have preferred us to have read his book. The choice, as usually happens, was not his.

WISTERIA

BOTANICAL NAME: *Wisteria*. FAMILY: *Leguminosae*.

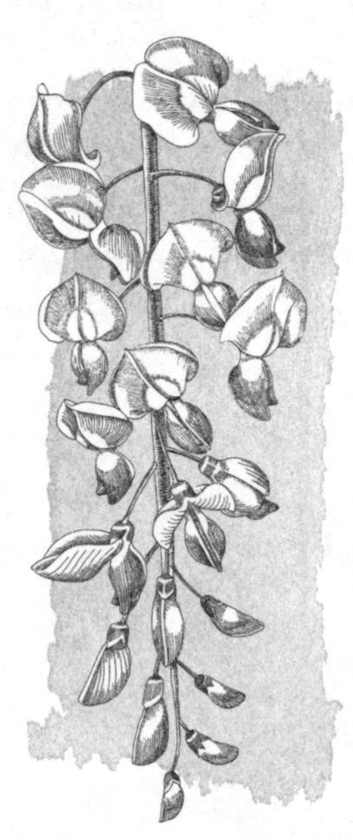

The wisteria is named for Doctor Caspar Wistar, who was a professor at the University of Pennsylvania, president of the Philosophical Society, and a distinguished botanist. He was a friend of President Jefferson, and, in 1795, had helped him identify some fossil remains of the giant sloth. In 1795, Jefferson had been sent bones of an enormous clawed animal, and he assumed that it was a kind of lion that might still be found somewhere in America. He called it a "megalonyx." Jefferson, and other Americans, were particularly defensive about large American animals because some Europeans, led by the French naturalist Georges Leclerc, comte de Buffon, maintained (partly to discourage immigration to America) that everything in America, including animals and people, was smaller and "degenerate" compared with Europe. Buffon had written *Histoire naturelle*, including thousands of facts and theories and comprising forty-four volumes, but he had no experience of America, where he said the air was so thin that the sun's influence could not be felt, causing everything there to shrink. Clearly it was nonsense, but even so, Jefferson hoped that finding the megalonyx would help to dispel this sort of calumny, and the animal would turn out to be as big as lions and tigers found outside America (Buffon had pointed out the small size of American lions and wild cats). But a similar fossil to Jefferson's was found in Paraguay and determined to be not a large cat but a giant sloth, now called a megathere. The struggle to gain European respect for American thought and civilization continued.

• Georges Leclerc, comte de Buffon, maintained that everything in America, including animals and people, was smaller and "degenerate" compared with Europe.

The wisteria we grow in our gardens is either from China *(Wisteria sinensis)* or Japan *(Wisteria floribunda)*, but there is a native American wisteria which was sent by Mark Catesby to England in 1724 as the "Carolina Kidney Bean." It was first given the botanical name of *Glycine frutescens* (the glycine family includes the soybean) and was grown in England, but not widely.

In 1818 John Reeves, who was a tea inspector for the East India Tea Company, sent a Chinese wisteria back to London. The famous Victorian garden writer John Loudon had tried to get the plant named *Consequa* because Reeves had obtained it from a Cantonese merchant of that name, who apparently died in poverty, unrecognized by the West. It is reassuring to know that he had been a successful swindler of English merchants in Canton. In 1818, the wisteria was named (misspelled) after Wistar by Thomas Nuttall, who came from England to Philadelphia and called America a country "full of hope and enthusiasm."

The Chinese wisteria was first thought to be tender and was planted in a hot greenhouse, where it nearly died. But by 1838, a plant growing outdoors in London had reached eleven feet in height and ninety by seventy feet in either direction. Ever since, wisterias have flourished both in America and Europe where stone walls and fences beg for their beauty. The American wisteria isn't nearly as large or fine as the Asian varieties—but, after all, what else can you expect from a degenerate American plant?

YARROW

COMMON NAMES: Yarrow, milfoil. BOTANICAL NAME: *Achillea*. FAMILY: *Compositae*.

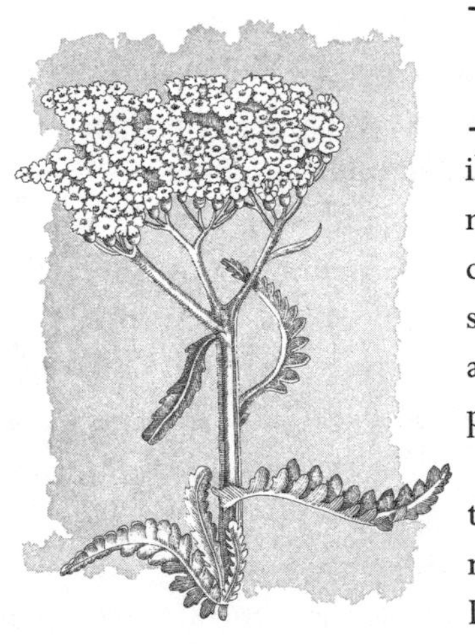

It's hard for us nowadays to understand what it must have been like to be ill in the past. We think of illness as having a physical cause and, mostly, a physical cure. If there is no cure we demand more medical research into it, confident that eventually one will be found. Not so in the past.

In the old days illness was thought to come from the stars, the mind, the humors, and sometimes the Devil. A plant that could heal was not only a medicine, it had mystical power as well. Yarrow was mainly used to stanch bleeding, and other names for it included "bloodwort," "stanchgrass," "sanguinary," and "woundewort." According to Dioscorides, the name "achillea" came from the fact that Achilles used it to heal his wounded Myrmidons, or soldiers, and he had been taught its secrets by Chiron the centaur, his tutor. Centaurs had the wisdom and hearts of men without their sexuality or slowness, combined with the swift power and other qualities of a horse. Achillea was thought to be particularly good for healing wounds made with iron, and so it was important to battlefield surgeons as late as the American Civil War, when the crushed plant was applied to bullet and shrapnel wounds.

But its power of healing was not only physical. The name "yarrow" comes from the Anglo-Saxon *gearwe*, the origin of which is uncertain, but some etymologists believe is from *gierwan*, "to prepare" or "to be ready." For yarrow was a defense against other ills. In France and Ireland it was one of the herbs of St. John that were picked (while they were still wet with dew) and burned on the eve of St. John's Day, to protect against evil. Often the fires were lit on the windward side of fields, so the protective smoke would blow over them, and burning brush would be carried around the stables. The French phrase *avoir toutes les herbes de Saint Jean* means "to be ready for anything." Nicholas Culpeper described "an ancient charm" whereby a leaf of yarrow was to be pulled off with the left hand as the sick man's name was pronounced.

Yarrow was brought to America with the earliest colonists and soon used by Native Americans as well as settlers. It spread fast and became a farmers' weed that cattle would not eat because of its bitter taste. The leaves contain tannin and make an astringent solution. It was also used for brewing beer. Surely it was effective and powerful, but it might have been overrated. In 1682 Abraham How, who kept a shop at Ipswich and wrote a book on healing, recommended combining yarrow with brandy and gunpowder for pain in the back as part of his method "to force Nature out of its own ordinary way." Such a medicine must have been potent—but perhaps not because of the yarrow.

YUCCA

COMMON NAMES: Yucca, Adam's needle, bear's grass. BOTANICAL NAME: *Yucca*. FAMILY: *Agavaceae*.

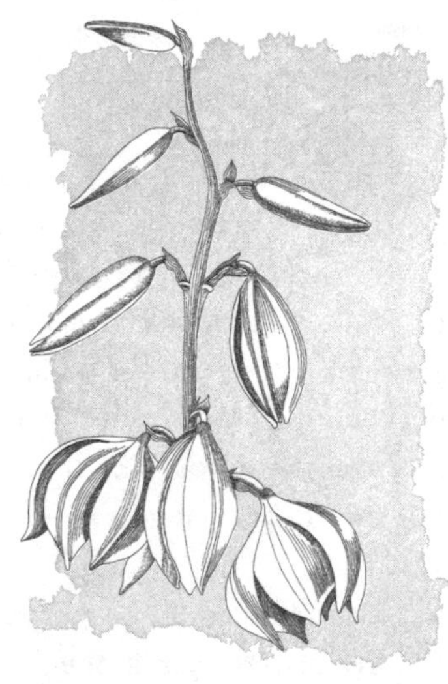

We are good at taking miracles for granted, but the pollination of the yucca has to astound us.

The yucca can only be fertilized by the female yucca moth, *Pronuba yuccasella*, who uses the plant exclusively to rear her young. Other insects feed on the nectar and pollen, but the yucca moth has another plan. When evening comes and the flowers are out, she collects pollen from the anthers and makes a sticky ball of it. After she has collected a ball a little larger than her head she wedges it under her chin and climbs the pistil of a different flower. Into this pistil she injects her eggs and then deposits her pollen ball on top of the stigma, rubbing and fixing it firmly. Thus, and only in this way, is the flower pollinated. In a few days the flowers wither and the moth larvae hatch. They eat some of the seeds in the pod and then drop to the ground, spin a cocoon, and later turn into adult moths. The remainder of the seeds make new plants.

With his customary meticulous observation, John Parkinson noticed that the yucca flowers dropped off after blooming "without bearing any seede in our Country, as farre as ever could be observed." He goes on to say that the yucca was first brought to England from the West Indies "by a servant of Master Thomas Edwards, an Apothecary of Exeter" and given to John Gerard. Gerard had named the plant "yucca," "supposing it to bee the true *Yuca* . . . wherewith the Indians make bread, called *Cassava*." Parkinson knew that it was not the cassava, or *Manihot esculenta*, but concluded that "not knowing by what better name to call it, let it hold still his first imposition, untill a fitter may be given it."

In fact it retained the name, although it was also known as "silk grass," because of the fibers that can be taken from the leaves, and "Adam's needle," because of its spiny leaves. Parkinson said he found the threads "so strong and hard" that they could not have been used for cloth. But William Byrd of Virginia wrote in 1728 that the Indians made aprons of it, which they "wear about their middles, for decency's sake." They also made ropes and baskets from the fibers. The fruits, called "datile," are said to be edible, and the roots contain saponin, which makes a soap-like lather.

The yucca became a popular garden plant that could be propagated by root cuttings, if not by seeds. It was a prized plant in the garden of Count Johann of Nassau at the Castle of Idstein. Count Johann employed Johann Walter of Strasbourg to make a *florilegium*, or flower list, and in it his yucca is illustrated with the proud note that on June 23, 1653, it bore 253 flowers. This garden fell into disrepair after the count died, and Idstein was destroyed by French troops in May 1795. Probably all that remains of the yucca is in the *florilegium*—one instance where the ancient and mysterious partnership that evolved between the yucca and the pronuba came to nothing.

ZINNIA

BOTANICAL NAME: *Zinnia*. FAMILY: *Compositae*.

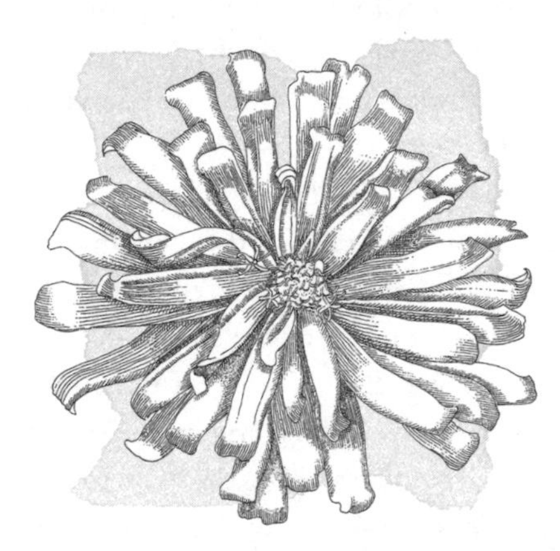

The zinnia, like the cosmos (from the Greek *cosmos*, "beautiful"), was sent from Mexico by Professor Casimir Gomez de Ortego to his friend the marchioness of Bute in Madrid.

The name "zinnia" comes from Johann Gottfried Zinn, who was a medical professor at Göttingen University. He wrote a description of the flora around Göttingen and, in 1753, he also published a book on the anatomy of the eye. He was the first to describe the iris of the eye in detail, a description that is accurate even now. He also discovered that the eyeball of a man is larger than that of a woman, regardless of their height. He is said to have written his book under severe but unspecified "domestic difficulties." Could his wife have been jealous of his passion for eyeballs? He died in 1759, aged only thirty-two, from a "most consumptive disease," and is remembered by a part of the eye called "Zinn's zonule," as well as our garden flower.

The zinnia in its native Mexico was called *mal de ojos* by the Spaniards, because the flowers were small and considered ugly to the eye. In fact, although they were named for Zinn by Linnaeus, nobody took much notice of them for two hundred years.

Zinn might have been pleased to see how nicely zinnias strike the eye these days. They were not really improved until this century, when one flower in a whole field of experimental zinnias, grown by Burpee, was used as the basis for breeding most of the hybrids we know. It was in the sixty-sixth row and known in the trade as "Old 66."

❧ The zinnia in its native Mexico was called *mal de ojos* by the Spaniards, because the flowers were small and considered ugly to the eye.

The marchioness of Bute who had sent the seeds to London was wife of the British ambassador to Madrid and daughter-in-law to John Stuart Bute (for whom the stewartia tree is named). Bute was the director of the Royal Botanic Gardens at Kew. When Frederick, Prince of Wales, died "by standing in the wet to see some trees planted" and getting pneumonia, his widow Augusta continued to supervise the Kew gardens closely. In 1785 Bute had published a nine-volume *Botanical Tables*, "composed solely for the amusement of the fair sex." One hopes they appreciated it. His admiration for the fair sex included Princess Augusta, and "as soon as the Prince was dead, they walked more and more, in honour of his memory."

Bute died an appropriate botanist's death. He fell off a cliff while reaching for a rare plant and never recovered from his injuries. In the language of flowers, zinnia stands for "thoughts of absent friends"!

FURTHER READING

Aden, Paul. *The Hosta Book* (Timber Press, 1988).

Alcock, Randal. *Botanical Names for English Readers* (Reeve & Co., 1876).

Allan, Mea. *Plants That Changed Our Gardens* (David & Charles, 1974).

Anderson, A. W. *The Coming of the Flowers* (Williams & Norgate, 1950).

Arkell, Reginald. *Old Herbaceous* (Harcourt Brace, 1951).

Bailey, L. H. *How Plants Get Their Names* (Dover, 1963).

———. *The Standard Cyclopedia of Horticulture* (Macmillan, 1922).

Bartram, John. *The Correspondence of John Bartram (1734–1777)*. Edited by Edmund and Dorothy Smith Berkeley (University Presses of Florida, 1992).

Bauman, Hellmut. *The Greek Plant World in Myth, Art, and Literature*. Translated by William T. and Edwyth Ruth Stearn (Timber Press, 1993).

Beale, Katherine. *Flower Lore* (Holt, 1917).

Bennett, Jennifer. *Lilies of the Hearth* (Camden House, 1991).

Blanchan, Neltje. *Nature's Garden* (Doubleday, 1905).

Blunt, Wilfrid. *The Art of Botanical Illustration* (Collins, 1971).

———. *The Compleat Naturalist: A Life of Linnaeus* (Collins, 1984).

———. *In for a Penny* (Hamish Hamilton, 1978).

Bretschneider, E. *A History of European Botanical Discoveries*. 2 vols. 1898 (Reprint, Zentral Antiquariet Der Deutschen Demokratischen Republic, 1981).

Briggs, Roy. *Chinese Wilson* (Royal Botanic Gardens, Kew, 1993).

Brosse, Jacques. *Great Voyages of Discovery: Circumnavigators and Scientists 1764–1843* (Facts on File, 1983).

Coats, Alice M. *Flowers and Their Histories* (Hulton Press, 1956).

———. *Garden Shrubs and Their Histories* (Vista Books, 1963).

———. *The Plant Hunters* (McGraw-Hill, 1969).

———. *The Treasury of Flowers* (McGraw-Hill, 1975).

Coats, Peter. *Flowers: The Story of Flowers, Plants and Gardens Through the Ages* (Weidenfeld & Nicolson, 1970).

Codd, L. E. "The South African Species of Kniphofia." In *Bothalia*, vol. 9 (Botanical Research Institute, South Africa, 1968).

Coffrey, Timothy. *North American Wildflowers* (Facts on File, 1993).

Coombes, Allen J. *Dictionary of Plant Names* (Timber Press, 1991).

Cowell, F. R. *The Garden as a Fine Art* (Houghton Mifflin, 1978).

Cox, E. H. M. *Plant Hunting in China* (Collins, 1945).

Cribb, Phillip, and Christopher Bailes. *Orchids* (Running Press, 1992).

Dana, Mrs. William Starr. *How to Know the Wild Flowers* (Charles Scribner, 1893).

D'Andrea, Jeanne. *Ancient Herbs in the Paul Getty Museum* (Paul Getty Museum, 1982).

Darwin, Charles. *Collected Papers*. Edited by Paul H. Barrett (University of Chicago Press, 1977).

———. *The Movements and Habits of Climbing Plants*. 1875 (Reprint, New York University Press, 1988).

———. *The Origin of the Species*. 1859 (Reprint, Oxford University Press, 1951).

———. *The Power of Movement in Plants* (Reprint, Da Capo Press, 1966).

Duthie, Ruth. *Florists' Flowers and Societies* (Shire Garden History, 1988).

Duval, Marguerite. *The King's Garden* (University Press of Virginia, 1982).

Earle, Alice Morse. *Old-Time Gardens* (Macmillan, 1901).

Ellacombe, Canon Henry N. *In a Gloucestershire Garden* (National Trust Classics, 1982).

———. *The Plant Lore and Garden Craft of Shakespeare* (Edward Arnold, 1896).

Evans, Howard Ensign. *Pioneer Naturalists* (Henry Holt, 1993).
Fell, Derek. *The Impressionist Garden* (Carol Southern Books, 1994).
Fisher, John. *The Origins of Garden Plants* (Constable, 1989).
Folkard, Richard. *Plant Lore and Legend* (Samson, Lows, Marston & Co., 1892).
Freeman, Margaret B. *The Unicorn Tapestries* (Metropolitan Museum of Art, 1983).
Gayley, Charles Mills. *Classic Myths*. 1893 (Reprint, Ginn & Co., 1939).
Gordon, Jean. *Pageant of the Rose* (Studio Publications, 1953).
Gorer, Richard. *The Development of Garden Flowers* (Eyre & Spottiswood, 1970).
Greene, Edward Lee. *Landmarks of Botanical History* (Stanford University Press, 1983).
Grigson, Geoffrey. *A Dictionary of English Plant Names* (Allen Lane, 1974).
———. *The Englishman's Flora* (Phoenix House, 1955).
Hadfield, Miles. *British Gardeners* (A. Zwemmer Ltd. & Condé Nast Publications, 1980).
———. *Pioneers in Gardening* (Routledge & Kegan Paul, 1955).
Hamilton, Edith. *Mythology* (Mentor Books, 1942).
Harvey, John. *Early Nurserymen* (Phillimore & Co., 1974).
Hawkes, Ellison. *Pioneers of Plant Study* (Sheldon Press, 1928).
Healey, B. J. *A Gardener's Guide to Plant Names* (Charles Scribner's Sons, 1972).
———. *The Plant Hunters* (Charles Scribner's Sons, 1975).
Hendrick, U. P. *History of Horticulture in America* (Timber Press, 1988).
Hendrickson, Robert. *Ladybugs, Tiger Lilies and Wallflowers* (Prentice Hall, 1993).
Hepper, Nigel. *Plant Hunting for Kew* (Royal Botanic Gardens, Kew, 1989).
Hill, Thomas. *The Gardener's Labyrinth, 1652*. Edited by Richard Mabey (Oxford University Press, 1987).
Hirschberg, Julius. *History of Ophthalmology*, vol. 5 (Bonn, 1985).
Hulton, Paul, and Lawrence Smith. *Flowers in Art from East and West* (British Museum Publications, 1979).

Hutchinson, John, and Ronald Melville. *The Story of Plants* (P. R. Gawthorne, 1948).

Huxley, Anthony. *Plant and Planet* (Viking Press, 1974).

Huxley, Elspeth. *Livingstone and His African Journey* (Saturday Review Press, 1974).

Hyam, Roger, and Richard Pankhurst. *Plants and Their Names* (Oxford University Press, 1995).

Kalm, Peter. *Travels in North America*. Edited by E. Benson (Dover, 1987).

Krastner, Joseph. *A Species of Eternity* (Alfred A. Knopf, 1977).

The Larousse Encyclopedia of Mythology. Edited by Felix Guirand and translated by Richard Aldington and Delano Ames (Paul Hamlyn, 1959).

Leighton, Ann. *American Gardens in the Eighteenth Century* (University of Massachusetts Press, 1986).

———. *American Gardens in the Nineteenth Century* (University of Massachusetts Press, 1986).

———. *Early American Gardens* (University of Massachusetts Press, 1986).

Lemmon, Kenneth. *The Covered Garden* (London Museum Press, 1962).

———. *The Golden Age of Plant Hunters* (Phoenix House, 1986).

Loudon, J. C. *An Encyclopaedia of Gardening* (Longman, Rees, Orme, Brown, Green, & Longman, 1834).

Lu, Henry C. *Legendary Chinese Healing Herbs* (Sterling, 1991).

Lyte, Charles. *The Plant Hunters* (Orbis, 1983).

Maurois, André. *A History of France* (Grove Press, 1960).

The Magic and Medicine of Plants (Reader's Digest, 1986).

McLaughlin, Terence. *Dirt: A Social History as Seen Through the Uses and Abuses of Dirt* (Stein and Day, 1971).

McLean, Teresa. *Medieval English Gardens* (Collins, 1981).

Moldenke, Harold, and Alma Moldenke. *Plants of the Bible* (Dover, 1952).

Monardes, Nicolas. *Joyfull Newes out of the Newe Founde Worlde*. Translated by John Frampton. 1577 (Facsimile, Constable, 1925).

Moorhead, Alan. *Darwin and the Beagle* (Penguin Books, 1971).

Neal, Bill. *Gardener's Latin* (Algonquin, 1992).

Nicholson, George. *Dictionary of Gardening* (L. Upcott Gill, 1887).

O'Brian, Patrick. *Joseph Banks* (David Godine, 1993).

Pankhurst, Alice. *Who Does Your Garden Grow?* (Earl's Eye, 1992).

Parkinson, John. *Garden of Pleasant Flowers*. 1629 (Reprint, Dover, 1976).

Paxton, Joseph. *Botanical Dictionary* (Bradbury Evans & Co., 1868).

Phillips, Henry. *History of Cultivated Vegetables* (Henry Colburn & Co., 1822).

———. *Sylva Florifera* (Longmans, Hurst, Rees, Orme and Brown, 1823).

Prest, John. *The Garden of Eden* (Yale University Press, 1981).

Randall, Vernon. *Wild Flowers in Literature* (Scholastic Press, 1934).

Reinikker, Merle. *A History of the Orchid* (University of Miami Press, 1972).

Reveal, James L. *Gentle Conquest* (Starwood, 1992).

Richardson, Rosamond. *Roses: A Celebration* (Piatkus, 1984).

Rohde, Eleanour Sinclair. *The Old English Herbals* (Dover, 1971).

Ruskin, John. *Proserpina* (Estes & Lariat, 1897).

Shaver, Claire Haughton. *Green Immigrants* (Harcourt Brace, 1978).

Smith, A. W. *A Gardener's Book of Plant Names* (Harper & Row, 1963).

Spongberg, Stephen. *A Reunion of Trees* (Harvard University Press, 1990).

Stafleu, Frans. A., and Richard S. Cowan. *Taxonomic Literature* (Bohn, Scheltema & Holkema, 1976).

Stearn, William T. *Stearn's Dictionary of Plant Names for Gardeners* (Cassell, 1972).

Steele, Arthur. *Flowers for the King* (Duke University Press, 1946).

Stockwell, Christine. *Nature's Pharmacy* (Royal Botanic Gardens, Kew, and Century, 1988).

Stuart, David. *The Garden Triumphant* (Harper and Row, 1988).

Sturtevant, Lewis. *Edible Plants of the World* (Dover, 1972).

Tergit, Gabriele. *Flowers Through the Ages* (Oswald Wolfe, 1961).

Thomas Jefferson's Garden Book. Edited by Edwin Betts (American Philosophical Society, 1944).

Waters, Michael. *The Garden in Victorian Literature* (Scolar Press, 1988).
Webber, Ronald. *The Early Horticulturalists* (Charles Newton Abbot, 1986).
Whittle, Tyler. *The Plant Hunters* (PAJ Publications, 1988).
Whittle, Tyler, and Christopher Cooke. *Curtis's Flower Garden Displayed* (Magma Books, 1991).
Wilson, E. H. *If I Were to Make a Garden* (Stratford, 1931).
———. *Plant Hunting* (Stratford, 1927).
Wright, Richardson. *Gardener's Tribute* (J. B. Lippincott, 1949).

INDEX